Essentials of

Nonlinear Circuit Dynamics with MATLAB® and Laboratory Experiments

T0141318

Essentials of
Nonlinear Circuit Dynamics with MATLAB® and Laboratory Experiments

Arturo Buscarino • Luigi Fortuna
Mattia Frasca

CRC Press
Taylor & Francis Group
Boca Raton London New York

CRC Press is an imprint of the
Taylor & Francis Group, an **informa** business

CRC Press
Taylor & Francis Group
6000 Broken Sound Parkway NW, Suite 300
Boca Raton, FL 33487-2742

First issued in paperback 2021

© 2017 by Taylor & Francis Group, LLC
CRC Press is an imprint of Taylor & Francis Group, an Informa business

No claim to original U.S. Government works

ISBN-13: 978-0-367-78222-1 (pbk)
ISBN-13: 978-1-138-19813-5 (hbk)

Library of Congress Cataloging-in-Publication Data

Names: Buscarino, Arturo, author. | Fortuna, L. (Luigi), 1953- author. | Frasca, Mattia, author.
Title: Essentials of nonlinear circuit dynamics with MATLAB and laboratory
   experiments / Arturo Buscarino, Luigi Fortuna, and Mattia Frasca.
Description: Boca Raton : Taylor & Francis, a CRC title, part of the Taylor & Francis
   imprint, a member of the Taylor & Francis Group, the academic division of T&F
   Informa, plc, [2017] | Includes bibliographical references and index.
Identifiers: LCCN 2016054358| ISBN 9781138198135 (hardback : acid-free paper) | ISBN
   9781315226309 (ebook : acid-free paper)
Subjects: LCSH: Electric circuits, Nonlinear-- Design and construction-- Data processing.
   | Electric circuits, Nonlinear-- Experiments. | MATLAB.
Classification: LCC TK454 .B87 2017 | DDC 621.3815-- dc23
LC record available at https://lccn.loc.gov/2016054358

Visit the Taylor & Francis Web site at
http://www.taylorandfrancis.com

and the CRC Press Web site at
http://www.crcpress.com

Contents

4 Oscillators   63

5 Strange attractors and continuous-time chaotic systems   107

6 Cellular nonlinear networks   151

Preface

Even if we often study real phenomena with linear or linearized models, in reality these are approximations of nonlinear systems. While for both analysis and control of linear systems, a unified theory has been developed over the years, the study of nonlinear systems requires ad hoc methods which are still subject of intense research. There are a lot of methodologies for the modeling, the analysis and control of nonlinear systems. Those that we have dealt with in this book have been intended to provide to students the essential details to approach the study of nonlinear system dynamics. These guidelines follow a route that introduces the reader to nonlinear systems and to the implementation of nonlinear circuits that can provide an added value with respect to linear ones. Therefore, the spirit of the student in studying this book should be to gain further possibilities in designing innovative systems and to make the nonlinear feature an active element of the system.

The book includes the guidelines of a course for the study of dynamical nonlinear circuits. The main elements to face the study of nonlinear dynamics are discussed following a route inspired by a modified version of the scheme proposed by Varela to classify complex systems. The various chapters and their contents are organized in the book following the same order adopted during the course, Complex Adaptive Systems, that the authors held at the University of Catania for several years, and where a particular emphasis is given to the topic of nonlinear dynamics of electronic circuits.

The book is organized in eight chapters. In the first chapter introductory elements of nonlinear dynamics are presented, outlining in particular the main differences between linear and nonlinear systems. In Chapter 2, by using the classical discrete-time logistic map, the main topics related to the nonlinear behavior of dynamical systems are dealt with. In the same chapter some essential concepts of chaos theory are also introduced. In the book great attention has been devoted to the bifurcation diagram as a tool to characterize the behavior of a system. In Chapter 3 the main concepts illustrated with simple examples are reported to face the problem of elementary, perfect, and imperfect bifurcations. The examples reported in this chapter are referred to maps and first-order systems. In Chapter 4, 2D maps and second-order autonomous systems are studied. In particular, some important bifurcations of 2D maps are discussed and, as concerns continuous-time second-order systems, their properties are illustrated through examples of canonical oscillators (van der Pol, Hewlett circuit) by also using analytical tools. In Chapter 5 the behavior of third-order systems with strange attractors is analysed. The concept of hyperchaos is also outlined. In Chapter 6 cellular nonlinear networks (CNNs) are presented. The state controlled CNNs are defined and the principal issues related to these nonlinear circuits are illustrated. The paradigm of CNNs allows us to design and realize simple useful electronic circuits governed by the same dynamics of canonical chaotic circuits. In Chapter 7 synchronization and control of chaotic circuits are discussed. In Chapter 8 laboratory experiments regarding simple electronic nonlinear circuits, chaotic networks, and their synchronization are included. Moreover, the hybrid (digital/analog) implementation of dynamical discrete-time system with the help of the Arduino® microcontroller has been introduced. In the same chapter, the innovative component, the memristor, has been proposed to realize new chaotic dynamics.

The book is therefore organized to offer a complete course presenting the main concepts of nonlinear dynamics by using electronic circuits. The essential theory is supported by numerical examples with MATLAB®. Moreover, laboratory experiments are included; they are structured in such a way that each concept is illustrated by a real experiment. In this way, the reader who wants to understand nonlinear dynamics has the possibility to acquire in a short time significant experience in the field. The book has been conceived after about 20 years of experience in teaching the topics. It also includes research results obtained by the authors in the field.

1

Introduction to nonlinear systems

CONTENTS

The first chapter of this book aims to introduce the readers to the basic notions of complex systems. Reviewing the classifications commonly adopted in literature, a new perspective over complex systems is proposed that allows us to discuss the concepts of nonlinearity, uncertainty, and system dimension unveiling the role that the interplay between them has in the emergence of complex behaviors. Furthermore, some numerical and analytical tools useful for the study of complex systems are introduced.

1.1 Classification of complex systems

The study of complex systems has fascinated people from a wide range of scientific fields, since these systems became, in the last decades, a well-established paradigm to describe both natural and artificial phenomena. In order to provide the necessary guidelines to study and classify complex systems, let us start considering the qualitative scheme reported in Figure 1.1. This outstanding diagram, originally proposed by Weinberg [93], classifies complex systems with respect to the level of uncertainty and complexity. The clusters that can be retrieved in such a diagram show an increasing level of disorder when uncertainty grows, ranging from ordered to random structures. The increase of the complexity level allows us to identify a region in the diagram characterized by the emergence of patterns and organization as a result of the interplay between complexity and uncertainty. This range is called homeody-

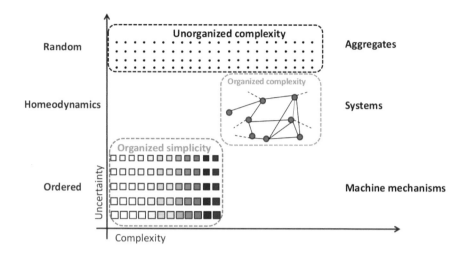

FIGURE 1.1
Systems at different levels of complexity and uncertainty.

namic and is characterized by the existence of processes where self-adaptation and self-organization play a crucial role.

Along with the conceptual diagram by Weinberg, an abstract representation and synthesis of complex systems can be schematized as in Figure 1.2 by means of a hierarchical aggregation graph, including also the dynamic relationship between each level, leading to emergence [24]. Complex systems, in fact, attain different levels of organization as size scales. It often occurs that the higher the number of elements the higher the level of organization, which may occur as a consequence of the self-organization process, or as the effect of a control strategy.

This book will present the different topics related to the dynamics of complex adaptive systems following a route through the scheme reported in Figure 1.3. This represents a modified version of the diagram first introduced by Varela et al. [91] and reported also by Schreiber [76]. The original diagram was meant to merge the two paradigms of nonlinear determinism and linear stochasticity in a unified framework, since they were considered in the '90s the only ones with a solid mathematical background. Hence, the two paradigms represent the extreme positions of the original diagram. The modified version reported in Figure 1.3 considers a further axis which takes into account the role of the number of state variables. This modification of the Varela diagram allows us to combine the classification presented by the diagrams in Figures 1.1 and 1.2. In fact, there is an intrinsic relationship among the three figures as they give a global view of complex systems classification.

The third dimension added in the Varela diagram allows us to introduce the concept of "networks of dynamical elements" which mirrors the organized complexity depicted in Figure 1.1. Therefore, in the following we will refer to the diagram in Figure 1.3 to introduce the various topics discussed in the book chapters. We will introduce the essential elements that allow us to establish a dichotomy between each element of Figure 1.3 and the various items reported in the book.

A further fundamental point is that, in this book, we aim to establish and exploit a strict link between dynamical complex systems, mathematical models, and electronic circuits and devices. Under this perspective, complex systems will be investigated on the basis of specifically designed electronic circuits and devices, which implement the relevant mathematical models and mimic their behavior. Working with electronic devices, in fact, often gives more insight than numerically computing model solutions, which are always subjected to an approximation error. Moreover, real circuits allow us to clearly understand the role of nonlinear interactions in the emergence of peculiar behaviors. The essential analytical methods will also be easily mapped from one domain to the other.

The element of universality between different domains is that complex systems encompass both nonlinearity and uncertainty which are always present in electronic devices. Furthermore, in the framework of complex systems both elements, rather than drawbacks, represent key factors for the emergence of robustness. As a consequence, despite the fact that nonlinearity and uncertainty make difficult the definition of a general theory to study such systems, they allow us to obtain reliable electronic setups showing complex behavior.

Both continuous-time and discrete-time complex systems will be considered. While for the first class, experiments can be done using implementation of electronic analogues, the experimentation with the latter class exploits the use of the increasing computational resources of microcontrollers, now available at low cost.

Consider now the origin of the graph reported in Figure 1.3. It represents the simplest class of systems, that is, first-order linear deterministic systems, i.e., linear systems with only one state variable and with constant and certain parameters. In the other extreme of the graph, the region of disorganized systems and circuits appears, that is, systems where uncertainty, nonlinearity, and the state variables are aggregates and not organized in structured, distributed cooperation. In the central region we observe organized structures where coordinated arrangements form and promote stability, robustness, and adaptation. In agreement with Figures 1.1 and 1.2, this model leads us to organized complexity and complex systems. In our perspective, they are mathematically represented by the reaction-diffusion equations with noise whose electronic analogues are reaction-diffusion cellular nonlinear networks (RD-CNNs), that is, lattices of electronic nonlinear circuits. Therefore, at the core of the graph the complex adaptive behavior of global level structures emerges as the result of the contribution of the nonlinearity of each single unit and of the spatiotemporal interactions.

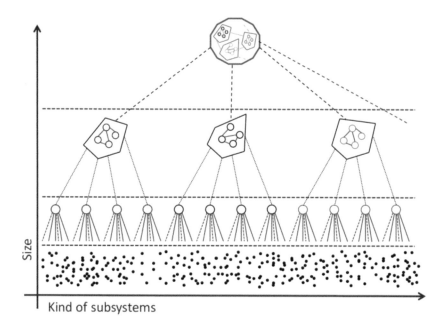

FIGURE 1.2
Hierarchical aggregation graph of complex systems.

1.2 First-order systems

In order to discuss the emergent properties of complex systems we will first start from the origin of the graph reported in Figure 1.3, moving then step by step along its three dimensions. In particular, we discuss here some examples of first-order systems, their equilibria, and stability properties.

Example 1.1 _____

Consider the system

$$\dot{x} = -x \tag{1.1}$$

that, assuming that x represents the voltage across the capacitor C, corresponds to the circuit in Figure 1.4 with $C = 1\text{F}$ and $R = 1\Omega$. The solution of Equation (1.1) in the time domain can be analytically found. It is given by $x(t) = e^{-t}x(0)$ where $x(0)$ is the initial condition of the system, that is the voltage across C at time $t = 0$.
The energy stored in the capacitor is given by $V(x) = \frac{1}{2}x^2$, this also represents a *Lyapunov function* for the system.
In fact, $V(x)$ is a positive definite function and $\dot{V}(x) = -x^2$ is negative definite, thus the system (1.1) is asymptotically stable.

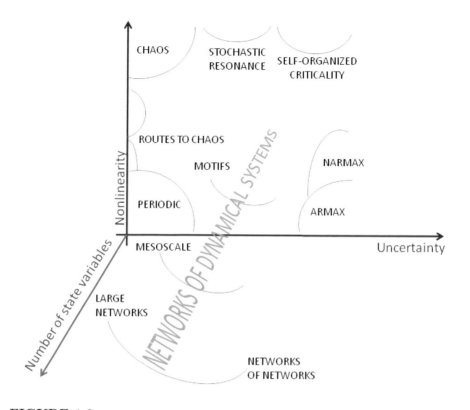

FIGURE 1.3
A modified state of art in complexity theory.

With the aim of making the reader familiar with MATLAB® code, as a first example we report here the commands to calculate and plot this Lyapunov function (the result is shown in Figure 1.5).

```
x = [-10:0.01:10];
y=0.5*x.^2;
plot(x,y,'k')
ylim([-10 50])
xlabel('x')
ylabel('V(x)')
```

Example 1.2 _____

Consider now the system

$$\dot{x} = x - x^3 \tag{1.2}$$

It represents a nonlinear first order dynamical system. In this case, the Lyapunov function is given by $V(x) = -\frac{1}{2}x^2 + \frac{1}{4}x^4$. The commands to plot it are listed here:

```
x = [-10:0.01:10];
y=-0.5*x.^2+(1/4)*x.^4;
```

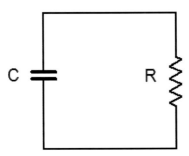

FIGURE 1.4
An RC circuit whose mathematical model can be written as the linear first-order system in Equation (1.1).

```
plot(x,y,'k')
ylim([-0.3 0.4])
xlabel('x')
ylabel('V(x)')
```

and the graph illustrated in Figure 1.6. $V(x)$ has two minima, corresponding to two stable equilibrium points. Depending on the initial condition, the system can reach one of the two states, so that we are in the presence of *bistability*.

1.2.1   Graphical analysis of equilibrium points

Given the system

$$\dot{x} = f(x) \tag{1.3}$$

with x scalar, and $f(x)$ representing the dynamics of the system, the intersection points between $f(x)$ and the x-axis defines all the equilibrium points of the system, i.e., $\dot{x} = 0$. The stability properties of these equilibrium points can be studied graphically: stability requires that in correspondence of $f(x) > 0$ the value of x will increase, while, when $f(x) < 0$, the value of x will decrease. So, an equilibrium point is stable if, for any small quantity $\varepsilon > 0$, $f(x - \varepsilon)$ is positive and $f(x+\varepsilon)$ is negative. Conversely, the equilibrium point is unstable.

The following commands and Figure 1.7 illustrate an example of a system with an infinite number of equilibrium points, i.e., $f(x) = \sin(x)$.

```
x=[-15:0.01:15];
y=sin(x);
plot(x,y,'k')
xlabel('x')
ylabel('f(x)')
ylim([-1.5 1.5])
xlim([-15 15])
```

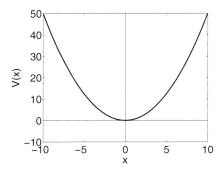

FIGURE 1.5
Lyapunov function for the RC circuit.

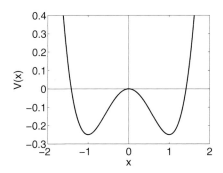

FIGURE 1.6
Lyapunov function for system (1.2).

```
hold on, plot([-15 15],[0 0],'k')
hold on, plot([0 0],[-1.5 1.5],'k')
hold on, plot([-4*pi -2*pi 0 2*pi 4*pi],[0 0 0 0 0],'ko')
hold on, plot([-3*pi -pi pi 3*pi],[0 0 0 0],'ko','MarkerFaceColor','k')
```

Considering again system (1.1), the graphical analysis of Figure 1.8 shows that for any initial condition the system reaches the origin that is the only equilibrium point. This point is stable.

If we consider, instead, system (1.2), the graphical analysis shown in Figure 1.9 reveals the existence of three equilibrium points: $x = \pm 1$ that are stable and $x = 0$ that is unstable. In this case, the final point reached by the system depends on the initial condition.

Note that linear systems may have an infinite number of equilibrium points (e.g., the system $\dot{x} = 0$) or, like system (1.1), a single equilibrium. If this point is asymptotically stable, any initial condition will drive the trajectory towards

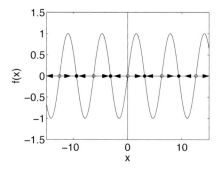

FIGURE 1.7
A system with an infinite number of equilibrium points.

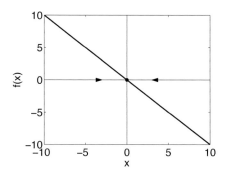

FIGURE 1.8
Equilibrium point for the linear system (1.1).

it. This means that only one piece of information is handled by using the stable equilibrium point of a linear system.

A quite different scenario appears in nonlinear systems where, in general, the system has more than one equilibrium point. This allows us to associate to each equilibrium point the set of initial conditions leading to it. This set is called the basin of attraction of the equilibrium.

Considering again Figure 1.7 we note that a basin of attraction is associated to each of the equilibrium points; thus, by exploiting the different equilibriums and basins of attractions, a nonlinear dynamical system, even if it has only one state variable, can handle more than one piece of information. A nonlinear system of order n has basins of attraction in an n-dimensional space associated to the equilibrium points in that space. Therefore, a nonlinear dynamical system could be a recipient of a rich variety of information. Taking into account that nonlinear systems also exhibit other types of attrac-

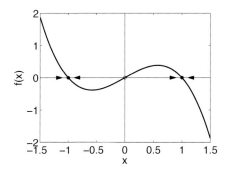

FIGURE 1.9
Equilibrium points for the nonlinear system (1.2).

tors (including limit cycles and strange attractors) the capability of nonlinear networks for handling information is considerable. From an engineering point of view the problem of designing nonlinear systems with given associative properties and selected dynamics is challenging. The design and implementation of the corresponding electronic circuits are a further central point of the realization of such nonlinear systems.

1.3 Numerical solutions of differential equations

In order to understand the emerging behavior that could arise in nonlinear systems let us discuss some mathematical essential concepts. Consider a linear time-invariant continuous-time system described by the following system of differential equations:

$$\dot{\mathbf{x}} = A\mathbf{x} \tag{1.4}$$

where $\mathbf{x} \in \mathbb{R}^n$ is the state vector and $A \in \mathbb{R}^{n \times n}$ a square matrix of constant, real coefficients. To simplify the problem, let us consider that the eigenvalues of the matrix A are distinct and, thus, A may be rewritten as: $A = T^{-1}\Lambda T$ where T is the eigenvector matrix (that is, the columns of T are the eigenvectors of A) and Λ is a diagonal matrix containing the eigenvalues of A. The solution $\mathbf{x}(t)$ is given by:

$$\mathbf{x}(t) = e^{At}\mathbf{x}(0) = T^{-1}e^{\Lambda t}T\mathbf{x}(0) \tag{1.5}$$

In the case of linear discrete-time systems the dynamics is:

$$\mathbf{x}(k+1) = \mathbf{F}\mathbf{x}(k) \tag{1.6}$$

where the $t = kT$ with $k = 0, 1, 2, \ldots$ and T is the discrete time interval. If system (1.6) derives from sampling a continuous-time process at given, regularly spaced, time instants, then $\mathbf{F} = e^{AT}$ with T being the sampling time. System (1.6) is a linear difference equation in vectorial form, and its solution is

$$\mathbf{x}(k+1) = \mathbf{F}^k \mathbf{x}(0) \tag{1.7}$$

We recall here that asymptotic stability is guaranteed by the conditions that all the eigenvalues of matrix A have negative real part for continuous-time systems, while for discrete-time systems the eigenvalues of matrix F must lie in the unit circle, i.e., have modulus less than one. Another peculiar difference between the two types of systems is that in the discrete-time domain there exists a class of systems that has input response equal to zero after a finite number of steps, while in the continuous-time case this condition cannot ever be met.

Once the initial conditions are fixed, the solutions of (1.4) or (1.6) may be expressed as the linear combination of simple modes, one for each of the eigenvalues of the matrix A or F. Accordingly, the global behavior is the sum of that of the elementary dynamics and emergent behavior cannot be expected.

In the case of nonlinear systems what happens may be totally different, as for these systems the behavior is often more than the sum of that of their single constituent parts. The general equation of a nonlinear system can be written as

$$\dot{\mathbf{x}} = f(\mathbf{x}) \tag{1.8}$$

with $\mathbf{x} \in \mathbb{R}^n$ and $f(\mathbf{x}) : \mathbb{R}^n \to \mathbb{R}^n$. Unlike what happens for linear systems, in general, system (1.8) cannot be analytically solved. In addition, in general, the solution cannot be decoupled as for linear systems. Therefore, the behavior of the system cannot be considered as equal to the sum of the behaviors of the single parts, and emergent behavior can occur.

To obtain a solution of system (1.8) with given initial conditions $\mathbf{x}(0)$, numerical methods like Runge–Kutta methods are often needed. We will describe one of them in the following section.

1.3.1   Runge–Kutta methods

Consider an initial value problem:

$$\begin{aligned}\dot{\mathbf{x}}(t) &= f(\mathbf{x}, t) \\ \mathbf{x}(t_0) &= \mathbf{x}(0)\end{aligned} \tag{1.9}$$

with $t \in \mathbb{R}^+$. Runge–Kutta methods are a family of numerical integration routines that allow us to compute the solution of (1.9) at times $t_k = kT$ where T is the integration step size [72]. We will shortly illustrate the fourth-order one based on a fixed integration step size. According to this method, given $\mathbf{x}_k$, the next sample $\mathbf{x}_{k+1}$ of the solution of (1.9) is calculated as:

$$\mathbf{x}_{k+1} = \mathbf{x}_k + \frac{1}{6}T(\mathbf{K}_0 + 2\mathbf{K}_1 + 2\mathbf{K}_2 + \mathbf{K}_3) \qquad (1.10)$$

with:

$$
\begin{aligned}
\mathbf{K}_0 &= f(\mathbf{x}_k, t_k) \\
\mathbf{K}_1 &= f(\mathbf{x}_k + \tfrac{T}{2}\mathbf{K}_0, t_k + \tfrac{T}{2}) \\
\mathbf{K}_2 &= f(\mathbf{x}_k + \tfrac{T}{2}\mathbf{K}_1, t_k + \tfrac{T}{2}) \\
\mathbf{K}_3 &= f(\mathbf{x}_k + T\mathbf{K}_2, t_k + T)
\end{aligned}
\qquad (1.11)
$$

The family of Runge–Kutta methods includes the Euler method. In this case, only the term $\mathbf{K}_0$, representing an increment equal to the slope of the function at the beginning of the interval, that is, in $\mathbf{x}_k$, is used. This method is faster than the Runge–Kutta but provides lower accuracy.

In both methods illustrated above, the step size T is considered constant. The family of Runge–Kutta methods also includes techniques based on a variable step size to allow for a better precision when required by the numerical solution of specific dynamics. Faster dynamics, in fact, may need smaller step size, while a larger step size can be adopted when slower dynamics occurs. The MATLAB® routine `ode45` implements a 4-th order Runge–Kutta method with variable step size. It can be used with the following syntax:

```
[t,x]=ode45(@fun,tspan,x0)
```

where `tspan=[t0 tf]` is the time interval in which the solution is computed, and `x0` the initial condition. The function returns the state variable vector computed for a series of samples (not uniform) in the interval `tspan` as given in the vector `t`. If a constant interval is required for the output data, then `tspan` has to be specified as `tspan=[t0:T:tf]`. Note that in this latter case, the step size adopted by the numerical routine is still variable, but numerical results are provided with a constant time interval T. The differential equations have to be specified in a file named `fun` (that has to be saved in the current directory of MATLAB® or in a directory included in the path) as illustrated in the following example for the system $\dot{x} = x - x^3$.

The system equations are defined in the file `examplenlsys.m` that contains the following lines:

```
function dxdt=examplenlsys(t,x)

dxdt=[x-x^3];
```

The solution for $x(0) = 0.01$ is calculated and displayed with the following commands:

```
[t,x]=ode45(@examplenlsys,[0 10],0.01);
figure,plot(t,x)
xlabel('t')
ylabel('x')
```

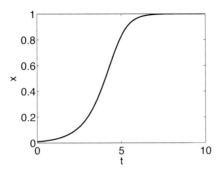

FIGURE 1.10
Solution of the system $\dot{x} = x - x^3$ with $x(0) = 0.01$.

The result is shown in Figure 1.10, which illustrates that the trajectory goes towards one of the two stable equilibrium points, i.e., $\bar{x} = 1$.

1.4   Exercises

1. What is a nonlinearity? Explain what you mean by nonlinearity. Include simple examples with physical meaning of nonlinear systems. Make a list of nonlinearities you meet during your daily life.

2. Let us consider the nonlinear system $\dot{x} = x^2 - 1$.

   (a) Find the analytical solution of the previous differential equation.
   (b) Find the numerical solution.
   (c) Compare the analytical and numerical solutions.
   (d) Propose a scheme with electronic devices representing the autonomous dynamical system discussed.

3. Consider the Gompertz law, which describes the growth of a tumor in terms of cells. Get more information on the internet, discuss the related models, and propose an electronic circuit that can emulate it.

4. Study the stability of the equilibrium points of the following systems:

   (a) $\dot{x} = \log x \sin x$;
   (b) $\dot{x} = x^2 - \log x$;
   (c) $\dot{x} = |x| + \log x$;

FIGURE 1.11
A mechanical system.

(d) $\dot{x} = \sin x / |x|$;

5. For each of the systems at the previous point find the potential function $V(x)$ and plot it.

6. Propose electronic circuits that have dynamical behavior governed by the equations at point 4.

7. Consider the mechanical system in Figure 1.11. Define a model of the type $\dot{x} = f(x)$ that approximately captures its behavior and calculate the potential function $V(x)$.

8. Consider the equation $m\dot{v} = mg - kv^2$ that was demonstrated by Carlson et al. [16] to be able to fit speed data from human skydivers. Discuss an interpretation of the formula and propose an electronic circuit emulating it.

9. With reference to Figures 1.1, 1.2, and 1.3, describe in 50 words their "ubiquity".

10. Make a list of the most impressive terms from this chapter and propose for each of them a mechanical/electronic device that can illustrate the concept.

11. Analyze the expression "define your terms, gentleman, define your terms. It saves argument!" Who pronounced it?

12. Propose simple experiments that show nonlinear phenomena in your everyday experience.

Further reading

For additional information on the topics of the chapter, the following references may be consulted: [24], [43], [55], [72], [91], [93].

2

The logistic map and elements of complex system dynamics

CONTENTS

The analysis of complex dynamical behaviors begins with the study of a paradigmatic system that is able to reveal a broad variety of nonlinear phenomena despite its particularly simple structure, i.e., the logistic map. We are thus moving near the origin of the Varela diagram, presented in Chapter 1, introducing a simple nonlinearity in a first-order system. The analysis of the logistic map is performed by means of numerical and analytical calculations introducing the reader to the concept of bifurcation.

2.1   The logistic map

Let us consider the discrete-time dynamical system

$$p_{i+1} = p_i + rp_i(1 - p_i) \tag{2.1}$$

System (2.1) is known as the Verhulst model [84]. It describes the size p_i of a population of living beings such as animals, organisms, bacteria, at discrete times $i = 0, 1, 2, \ldots$.

The model is derived taking into account the following considerations. Assume that a population grows at a rate r, thus

$$\frac{p_{i+1} - p_i}{p_i} = r \tag{2.2}$$

This equation represents a linear growth for the population. Verhulst realized that this linear model is unrealistic due to the fact that real populations do not indefinitely increase. Including environmental constraints, and thus limiting the growth, yields to model (2.1), where the nonlinearity incorporates a quadratic term that opposes the growth rate.

Equation (2.1) computes the population at time $i + 1$ considering the population at the previous time i, the linear contribution of the growth rate rp_i and its competitor, modeled as rp_i^2. Normalizing the equation with respect to the maximum value of the population size leads to the following map:

$$x_{i+1} = ax_i(1 - x_i) \tag{2.3}$$

The variable x_i is such that $0 \leq x_i \leq 1$. Equation (2.3) is known as the logistic map. It is a discrete-time dynamical system with only one state variable, a single parameter, a, and an elementary quadratic nonlinearity. For these reasons, the logistic map represents a structurally very simple dynamical system. However, simple dynamical systems do not necessarily lead to simple dynamical behaviors, and the logistic map is a paradigmatic example in this sense.

Assume that $y = x_{i+1}$ and $x = x_i$ so that

$$y = ax(1 - x) \tag{2.4}$$

Equation (2.4) highlights that the logistic map has the shape of a parabola intersecting the x-axis at points 0 and 1, with a vertex in $(\frac{1}{2}, \frac{a}{4})$. Given that x has to satisfy the constraint $0 \leq x \leq 1$, it results that $0 \leq a \leq 4$.

Expression (2.4) represents a family of logistic functions, that correspond to different parabolas obtained varying the parameter a.

Example 2.1 _____

By using the following MATLAB® commands we can find the family of logistic functions as in Equation (2.4):

```
hold on
x=linspace (0,1,101);
plot(x,x.*(1-x),'k.-','linewidth',2)
plot(x,2*x.*(1-x),'k--','linewidth',2)
plot(x,3*x.*(1-x),'k:','linewidth',2)
plot(x,4*x.*(1-x),'k','linewidth',2)
xlabel('x_i')
ylabel('x_{i+1}')
legend('a=1','a=2','a=3','a=4')
xlabel('x_i')
ylabel('x_{i+1}')
```

The result is shown in Figure 2.1.

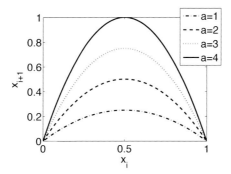

FIGURE 2.1
Family of logistic functions for different values of the parameter a.

Example 2.2 _____

Calculate for the logistic map the evolution of the state variable x_i.
Solution. To this aim, we use a series of commands that allow us to obtain the cobweb plot of the map, namely a graph where the iterations of the map are superimposed to the parabola, giving a graphical aid to visualize how the successive points are calculated from the previous ones. In fact, given the previous point $(x_{i-i}, x_i = f(x_{i-1}))$, one can go to the point (x_i, x_i) by following a horizontal line from $(x_{i-i}, x_i = f(x_{i-1}))$ to the straight line $y = x$, then one can follow a vertical line from (x_i, x_i) to $(x_i, f(x_i))$ to calculate the new point and iterate. The cobweb plot is built with the following commands illustrated for $a = 2.5$:

```
a=2.5;
n=100;
x=zeros(n+1,1);
x(1)=0.11;
for i=1:n, x(i+1)=a*x(i)*(1-x(i)); end
xx=linspace(0,1,101);
plot(xx,a*xx.*(1-xx),'r','linewidth',2)
hold on, plot([0 1],[0 1],'k','linewidth',2)
nstep=100; for i=1:nstep,plot(x(i)*[1 1],x(i:i+1),'b',...
x(i:i+1),x(i+1)*[1 1],'b','linewidth',2), end
xlabel('x_i')
ylabel('x_{i+1}')
```

In Figure 2.2(a),(b), and (c) the first, the second, and the n-th step of the cobweb plot construction are shown. To obtain the three plots, one has to vary the value of `nstep` in the previous set of commands.

2.2 Equilibrium points and periodic solutions of the logistic map

Let us compute the equilibrium points for the logistic map. An equilibrium point (also called a fixed point) $\bar{x}$ has to satisfy:

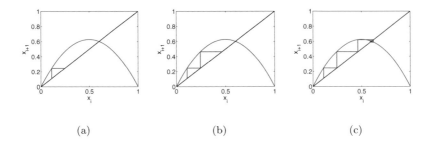

(a)   (b)   (c)

FIGURE 2.2
Construction of the cobweb plot for the logistic function with $a = 2.5$: (a) *nstep* = 1; (b) *nstep* = 2; (c) *nstep* = 100.

$$x_{i+1} = x_i = \bar{x} \tag{2.5}$$

Therefore, for the logistic map one gets:

$$\bar{x} = a\bar{x}(1 - \bar{x}) \tag{2.6}$$

and thus for $a \neq 0$

$$\bar{x} = \frac{a - 1}{a} \tag{2.7}$$

and

$$\bar{x} = 0 \tag{2.8}$$

The local stability of the equilibrium points is analyzed by computing the Jacobian of Equation (2.3):

$$\frac{dx_{i+1}}{dx_i} = (a - 2ax_i)|_{x_i = \bar{x}} \tag{2.9}$$

For $\bar{x} = 0$, one obtains $\frac{dx_{i+1}}{dx_i} = (a - 2ax_i)|_{x_i = \bar{x} = 0} = a$, that is the origin is stable for $|a| < 1$, while for $\bar{x} = \frac{a-1}{a}$ one gets $\frac{dx_{i+1}}{dx_i} = (a - 2ax_i)|_{x_i = \bar{x} = \frac{a-1}{a}} = 2 - a$. So, $\bar{x} = \frac{a-1}{a}$ is stable if $|2 - a| < 1$, that is $1 < a < 3$. In conclusion for $a < 3$, there is always exactly one equilibrium that is stable, and, as it is unique, it is also globally stable.

Example 2.3 _____

Consider the logistic map with $a = 2.3$ and initial condition $x(0) = 0.11$, find the trajectory and check that it converges to the equilibrium point.
Solution. The same code as in Example 2.2 can be used, provided that the parameter is fixed to $a = 2.3$. For this value of the parameter, one obtains

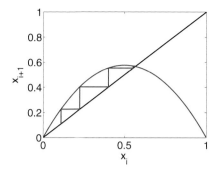

FIGURE 2.3
Cobweb plot of the logistic function for $a = 2.3$.

$\bar{x} = \frac{a-1}{a} = 0.5652$ (stable fixed point as $a < 3$). Figure 2.3 shows that in fact the trajectory converges towards this equilibrium.

Example 2.4 _____

Repeat the previous example with $a = 3.1$.
Solution. In this case, we consider the following commands:

```
a=3.1;
n=100;
x=zeros(n+1,1);
x(1)=0.11;
for i=1:n, x(i+1)=a*x(i)*(1-x(i)); end
xx=linspace(0,1,101);
plot(xx,a*xx.*(1-xx),'r','linewidth',2)
hold on, plot([0 1],[0 1],'k','linewidth',2)
nstep=100; for i=1:nstep,plot(x(i)*[1 1],x(i:i+1),'b',...
x(i:i+1),x(i+1)*[1 1],'b','linewidth',2), end
xlabel('x_i')
ylabel('x_{i+1}')
```

The result is shown in Figure 2.4(a), illustrating the first 100 steps of the evolution of the logistic map. As the transient is longer than 100 steps, we now readapt the MATLAB® commands to make a longer simulation and display the last 100 steps of it:

```
a=3.1;
n=1000;
x=zeros(n+1,1);
x(1)=0.11;
for i=1:n, x(i+1)=a*x(i)*(1-x(i)); end
xx=linspace(0,1,101);
plot(xx,a*xx.*(1-xx),'r','linewidth',2)
hold on, plot([0 1],[0 1],'k','linewidth',2)
nstep=1000; for i=901:nstep,plot(x(i)*[1 1],x(i:i+1),'b',...
x(i:i+1),x(i+1)*[1 1],'b','linewidth',2), end
xlabel('x_i')
ylabel('x_{i+1}')
```

The cobweb plot, shown in Figure 2.4(b), reveals that after a transient the system reaches a steady-state where the trajectory oscillates between two values $x_1 = 0.7646$ and $x_2 = 0.5580$. Hence, for $a = 3.1$ a period-2 cycle is obtained.

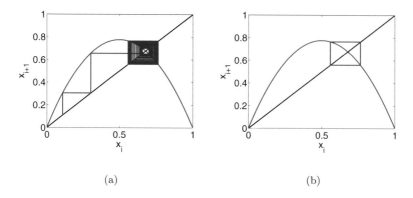

(a)　　　　　　　　　　　　　　　　(b)

FIGURE 2.4
Cobweb plot of the logistic function for $a = 3.1$: (a) transient; (b) steady-state.

We now find the conditions under which a period-2 solution exists. Let x_1 and x_2 be the two steady state values of x_i in the cycle, then it holds that:

$$x_2 = ax_1(1 - x_1)$$
$$x_1 = ax_2(1 - x_2)$$
(2.10)

By replacing the first equation into the second one and reordering the terms, we obtain:

$$ax_1^4 - 2a^3x_1^3 + (a^3 + a^2)x_1^2 + (1 - a^2)x_1 = 0$$
(2.11)

This expression is factorized as follows:

$$(a^2x_1^2 - a^2x_1 - ax_1 + 1 + a)(ax_1 + 1 - a)x_1 = 0$$
(2.12)

The solutions of Equation (2.12) are

$$x_1 = 0$$
$$x_1 = \frac{a-1}{a}$$
(2.13)

that are the equilibrium points previously obtained (in fact, they satisfy Equation (2.10) with $x_1 = x_2$) and the solutions of

$$a^2x_1^2 - a^2x_1 - ax_1 + 1 + a = 0$$
(2.14)

This is a second-order polynomial, whose discriminant is given by:

$$\Delta = (a^2 + a)^2 - 4a^2(1 + a)$$
(2.15)

If $\Delta > 0$, then two real solutions are obtained. The limit case $\Delta = 0$ gives

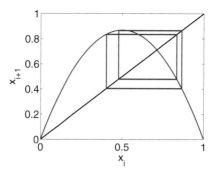

FIGURE 2.5
Cobweb plot of the logistic function for $a = 3.4695$.

the limit condition for the birth of period-2 cycles. In fact, $\Delta = 0$ for $a = 3$. For values of the parameter $a > 3$ there is, indeed, a window where the logistic map shows a bistable behavior.

Example 2.5 _____

Find by using MATLAB® the trajectory of the logistic map for $a = 3.4695$.
Solution. The following commands may be used:

```
a=3.4695;
n=10000;
x=zeros(n+1,1);
x(1)=0.11;
for i=1:n, x(i+1)=a*x(i)*(1-x(i)); end
xx=linspace(0,1,101);
plot(xx,a*xx.*(1-xx),'r','linewidth',2)
hold on, plot([0 1],[0 1],'k','linewidth',2)
nstep=10000; for i=9001:nstep,plot(x(i)*[1 1],x(i:i+1),'b',...
x(i:i+1),x(i+1)*[1 1],'b','linewidth',2), end
xlabel('x_i')
ylabel('x_{i+1}')
```

Note that the logistic map is simulated for $n = 10000$ steps, while only the last 1000 are visualized in the cobweb plot (shown in Figure 2.5). A closer inspection of it reveals a period-4 cycle.
Figure 2.6 reports the last 1000 steps of the evolution of the state variable, showing that the trajectory periodically jumps between four values, namely x_1, x_2, x_3, and x_4.

Summarizing the result of the previous examples, increasing the value of the parameter a from the condition where one stable equilibrium point exists, a period-2 cycle is observed and, then, a cycle with doubled periodicity, i.e., period-4, has been observed. For higher values of a the period further doubles, leading to a period-8 cycle and cycles with higher periodicity. This phenomenon, known as period doubling, is common to many nonlinear systems and often delineates a route to chaos known as the period doubling route to chaos.

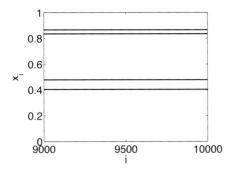

FIGURE 2.6
Evolution of the state variable for $a = 3.4695$, for which a period-4 cycle is observed.

2.3   Chaos in the logistic map

Despite its structural mathematical simple formulation, the logistic map is able to generate complex phenomena as the next example shows.

Example 2.6 _____

Find by using MATLAB® the trajectory of the logistic map for $a = 4$ and starting from the initial condition $x(0) = 0.11$. Then, compare the evolution of the system with that obtained for $x_0 = 0.11 + \varepsilon$ with $\varepsilon = 10^{-5}$.

Solution. The trajectory is obtained by using the MATLAB® commands:

```
a=4;
n=100;
x=zeros(n+1,1);
x(1)=0.11;
for i=1:n, x(i+1)=a*x(i)*(1-x(i)); end
xx=linspace(0,1,101);
plot(xx,a*xx.*(1-xx),'r','linewidth',2)
hold on, plot([0 1],[0 1],'k','linewidth',2)
nstep=n; for i=1:nstep,plot(x(i)*[1 1],x(i:i+1),'b',...
x(i:i+1),x(i+1)*[1 1],'b','linewidth',2), end
xlabel('x_i')
ylabel('x_{i+1}')
```

The result is shown in Figure 2.7, where we observe that the behavior of the map is not periodic. Each point in the diagram is different from the others. The map is iterated $n = 200$ times, but for any value of n the same structure is obtained. Consider now two trajectories starting from very similar initial conditions:

```
a=4;
n=200;
x1=zeros(n+1,1);x2=x1;
x1(1)=0.11;x2(1)=0.11+0.00001;
for i=1:n, x1(i+1)=a*x1(i)*(1-x1(i)); x2(i+1)=a*x2(i)*(1-x2(i)); end
figure,plot(x1), hold on,plot(x2)
figure,plot(x1), hold on,plot(x2,'r')
ylabel('x_{i}')
xlabel('i')
```

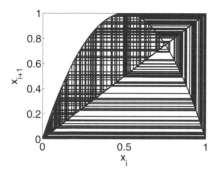

FIGURE 2.7
Chaos in the logistic map ($a = 4$).

As shown in Figure 2.8, the effect of a small change in the initial condition ($\varepsilon = 10^{-5}$) results in a trajectory $x2_i$ which is initially close to the one, $x1_i$, obtained with the previous initial condition but soon diverges to a totally different sequence of values, still maintaining the aperiodic characteristic. In order to evaluate such distinction between the two trajectories, the difference $x1_i - x2_i$ can be calculated. Although the initial conditions are quite similar, the difference is quite small only for the first iterations and then fluctuates aperiodically.

The high sensitivity to initial conditions is a prerogative of chaotic systems. Note that, if the same experiment is repeated for instance for $a = 2.3$, a totally different scenario is obtained, as after some iterations the error becomes zero (recall that for $a = 2.3$ the steady-state solution is a stable equilibrium point).

In linear systems, initial conditions only affect the transient behavior, while they have a negligible effect for the steady state. In nonlinear systems, as we have discussed for the logistic map with $a = 4$, the opposite may hold. Indeed for $a = 4$ two particular phenomena have been observed:

• the aperiodicity of the time series;

• the high sensitivity to initial conditions.

These two characteristics make the time series, which is generated by a deterministic system, unpredictable in the long term. For these reasons the time series generated by system (2.3) with $a = 4$ is said to have a chaotic behavior.

Exercise 2.1 _____

By using the **rand** function in MATLAB® generate a random series and compare it with the chaotic time series shown in Figure 2.8.
Solution. The commands

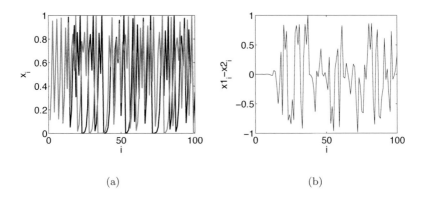

(a)   (b)

FIGURE 2.8
(a) Evolution of two chaotic signals generated by the logistic map ($a = 4$) with slightly different initial conditions and (b) their difference.

```
y=rand(100,1);
figure,plot(y,'k')
xlabel('i')
ylabel('y_i')
```

allow us to generate a stochastic time series with samples drawn randomly with uniform probability distribution in $[0, 1]$. The trend of the time series, shown in Figure 2.9, is quite similar to the chaotic one of Figure 2.8, as both are aperiodic and unpredictable. However, the two trajectories are generated in a totally different way; in one case the model is deterministic, in the other stochastic.

2.4 Intermittency

Intermittency is a peculiar deterministic phenomenon arising in nonlinear dynamical systems, where a sudden transition from an almost regular behavior to a chaotic one and vice versa occurs (other types of intermittency also exist, but for our purposes we will restrict the discussion only to this scenario).

According to [70], the phenomenon is described as "the dramatic interplay between bursts of chaos and almost periodic behavior." Intermittency includes the phenomenon of transient chaos, i.e., the case in which a chaotic trajectory eventually falls in a periodic orbit, on a fixed point, or diverges towards infinity.

In the logistic map, odd periodicity characterizes intermittency phenomena. Moreover, intermittency depends on initial conditions.

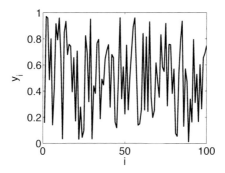

FIGURE 2.9
A stochastic signal.

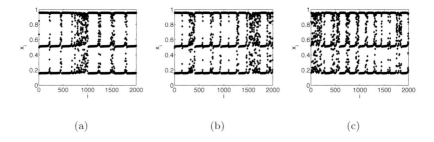

(a)   (b)   (c)

FIGURE 2.10
Intermittent behavior in the logistic map for $a = 1 + \sqrt{8} + \varepsilon$ with (a) $\varepsilon = -2.7 \cdot 10^{-5}$, (b) $\varepsilon = -5 \cdot 10^{-5}$, or (c) $\varepsilon = -7 \cdot 10^{-5}$.

Example 2.7 _____

In the logistic map, a period-3 cycle is obtained for $a = 1 + \sqrt{8}$, This value of the parameter is unique, as no other values of the parameter yield period-3 solutions. A proof of the uniqueness of the period-3 is given in the discussion of Exercise 2.2. Slightly decreasing the parameter a below $1 + \sqrt{8}$, intermittency appears. For instance, for $a = 1 + \sqrt{8} + \varepsilon$ with $\varepsilon = -2.7 \cdot 10^{-5}$, $\varepsilon = -5 \cdot 10^{-5}$ or $\varepsilon = -7 \cdot 10^{-5}$ the behavior of the logistic map is the one reported in Figure 2.10.

Example 2.8 _____

A period-5 cycle is obtained for $a = 3.906$. Intermittency appears for $a = 3.906 + \varepsilon$ with $\varepsilon = 7.7 \cdot 10^{-4}$ and $\varepsilon = 8 \cdot 10^{-4}$ as reported in Figure 2.11. Another period-5 cycle is obtained for $a = 3.73817237$ and intermittency, shown in Figure 2.12, appears for $a = 3.73817237 + \varepsilon$ with $\varepsilon = -2.4 \cdot 10^{-6}$.

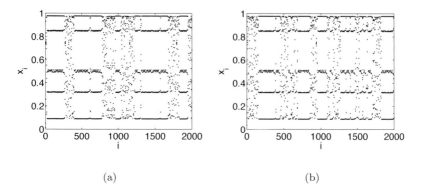

(a)   (b)

FIGURE 2.11
Intermittent behavior in the logistic map for $a = 3.906 + \varepsilon$ with (a) $\varepsilon = 7.7 \cdot 10^{-4}$ and (b) $\varepsilon = 8 \cdot 10^{-4}$.

Example 2.9 ───

A period-7 cycle is obtained for $a = 3.9222$. Intermittency appears for $a = 3.9222 + \varepsilon$ with $\varepsilon = 5.4 \cdot 10^{-4}$, as shown in Figure 2.13(a). Another period-7 orbit is obtained for $a = 3.70164076$ and intermittency, reported in Figure 2.13(b) occurs for $a = 3.70164076 + \varepsilon$ with $\varepsilon = -7.6 \cdot 10^{-7}$.

The following consideration can be made on period-odd solutions. Period-odd orbits are very sensitive to noise in the parameter, that can suddenly lead to intermittency. From a dynamical point of view, intermittency is characterized by dramatic changes yielding unpredictability of the behavior.

Exercise 2.2 ───

Find the condition for period-3 solutions in the logistic map and prove that this occurs for $a = 1 + \sqrt{8}$. Use MATLAB® and the symbolic toolbox.
Solution. Following an approach similar to that discussed for the search of period-2 solutions, we iterate the logistic map for searching a period-3 cycle:

$$x_1 = ax_3(1 - x_3) \tag{2.16}$$

$$x_2 = ax_1(1 - x_1) \tag{2.17}$$

$$x_3 = ax_2(1 - x_2) \tag{2.18}$$

where x_1, x_2, and x_3 are the periodic points of the map. Substituting Equation (2.16) in Equations (2.17) and (2.18) the following equation can be derived:

$$x_3 = -a^3 x_3 (a^2 x_3(ax_3(x_3 - 1) + 1)(x_3 - 1) + 1) \cdot \\ \cdot (ax_3(x_3 - 1) + 1)(x_3 - 1) \tag{2.19}$$

Now we define the polynomial F as:

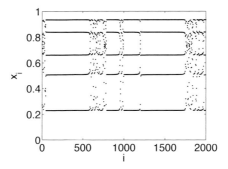

FIGURE 2.12
Intermittent behavior in the logistic map for $a = 3.73817237 + \varepsilon$ with $\varepsilon = -2.4 \cdot 10^{-6}$.

$$F = x_3 + a^3 x_3 (a^2 x_3 (a x_3 (x_3 - 1) + 1)(x_3 - 1) + 1) \cdot$$
$$\cdot (a x_3 (x_3 - 1) + 1)(x_3 - 1) \tag{2.20}$$

which identifies a family of curves depending on a as reported in Figure 2.14.
The condition that allows us to obtain the period-3 cycle is shown in Figure 2.14 where the three points of the period-3 cycle are marked. This condition consists in the fact that the iterator can assume only three values, namely the values corresponding to the markers in Figure 2.14, i.e., the tangent points of F with the x-axis.

In order to derive the analytical condition to obtain a, function F has to be divided for the quantity $x_3(x_3 - \frac{a-1}{a})$ since $x_3 = 0$ and $x_3 = \frac{a-1}{a}$ are the two fixed points of the quadratic iterator. The quotient q of this polynomial division can be calculated as:

$$\begin{aligned} q = {} & a^7 x_3^6 + \\ & + (-3a^7 - a^6) x_3^5 + \\ & + (3a^7 + 4a^6 + a^5) x_3^4 + \\ & + (-a^7 - 5a^6 - 3a^5 - a^4) x_3^3 + \\ & + (2a^6 + 3a^5 + 3a^4 + a^3) x_3^2 + \\ & + (-a^5 - 2a^4 - 2a^3 - a^2) x_3 + \\ & + a^3 + a^2 + a \end{aligned} \tag{2.21}$$

The existence of a period-3 cycle can be found considering that the parametric polynomial q must be divisible without remainder by the polynomial $F_1 = (x_3^3 + A x_3^2 + B x_3 + C)^2$, ensuring that solutions are three with double multiplicity. The solutions of q are the possible values of the logistic map for the given parameter value. Since the two trivial solutions corresponding to fixed points have been already removed, only three solutions can be obtained. Therefore, the parametric remainder R of the polynomial division $\frac{q}{F_1}$ can be equated to zero:

$$R = c_1 x_3^5 + c_2 x_3^4 + c_3 x_3^3 + c_4 x_3^2 + c_5 x_3 + c_6 = 0 \tag{2.22}$$

where the coefficient of the polynomial R are:

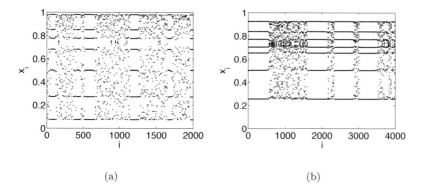

(a)   (b)

FIGURE 2.13

Intermittent behavior in the logistic map for (a) $a = 3.9222 + \varepsilon$ with $\varepsilon = 5.4 \cdot 10^{-4}$ and (b) $a = 3.70164076 + \varepsilon$ with $\varepsilon = -7.6 \cdot 10^{-7}$.

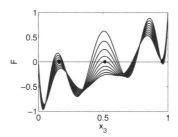

FIGURE 2.14

Condition to get a period-3 cycle: the red curve is obtained for $a = 1 + \sqrt{8}$, the three points of the cycle are marked in red.

$$c_1 = -2Aa^7 - a^6 - 3a^7$$
$$c_2 = a^5 + 4a^6 + 3a^7 - a^7(A^2 + 2B)$$
$$c_3 = -a^7(2C + 2AB) - a^4 - 3a^5 - 5a^6 - a^7$$
$$c_4 = (-B^2 - 2AC)a^7 + 2a^6 + 3a^5 + 3a^4 + a^3 \quad (2.23)$$
$$c_5 = -2BCa^7 - a^5 - 2a^4 - 2a^3 - a^2$$
$$c_6 = -C^2a^7 + a^3 + a^2 + a$$

Posing $c_1 = 0$, we derive $A = -\frac{3a+1}{2a}$. Substituting the calculated value in $c_2 = 0$, the second coefficient is $B = \frac{3a^2+10a+3}{8a^2}$. The third coefficient is then obtained substituting A and B in $c_3 = 0$ deriving $C = \frac{a^3-7a^2-5a-5}{16a^3}$. The three remaining polynomials, i.e., c_4, c_5, and c_6, are only in a, hence, if there exist common solutions inside the allowed range $[0, 4]$, a period-3 motion can be observed for those values of a. Among the possible solutions, there are only two of them which are common to the three polynomials, i.e., $a_1 = 1 + \sqrt{8}$ and $a_2 = 1 - \sqrt{8}$. Notice that $a_2 \approx -1.8284$ falls outside the allowed range of values for a and, consequently, has to be discarded.

Thus, there is a single value $a = a_1 = 1 + \sqrt{8}$ which leads to a period-3 trajectory. This assesses the uniqueness of the obtained value.

In order to perform these calculations in MATLAB®, we use the symbolic toolbox as shown in the following code:

```
syms a A B C x1 x2 x3
x1=a*x3*(1-x3);
x2=a*x1*(1-x1);

F=x3+a^3*x3*(a^2*x3*(a*x3*(x3-1)+1)*(x3-1)+1)*(a*x3*(x3-1)+1)*(x3-1);

X=(a-1)/a;

F1=x3*(x3-X);

[quot1,rem1]=quorem(F,F1);

f1=(x3^3+A*x3^2+B*x3+C)^2;

[quot2,rem2]=quorem(quot1,f1);
```

The function `syms` allows us to define the symbolic variable so that writing an expression involving symbols gives as output a further symbolic expression. Thus, F and $F1$ represent symbolic polynomials in $x3$. The function `quorem` realizes a polynomial division returning both quotient and reminder. The first division between F and $F1$ allows us to reduce polynomial F isolating the two known roots $x3 = 0$ and $x3 = \frac{a-1}{a}$ resulting in a quotient $quot1$ and a null reminder $rem1$. Since $f1$ is a generic polynomial with three solutions with double multiplicity, the polynomial division between $quot1$ and $f1$ according to the discussion outlined above, must return a zero remainder. We find that:

```
rem2=(-2*A*a^7-a^6-3a^7)*x3^5+(a^5+4*a^6+3*a^7-a^7*(A^2+2*B))*x3^4+
+(-a^7*(2*C+2*A*B)-a^4-3*a^5-5*a^6-a^7)*x3^3+((-B^2-2*A*C)*a^7+
+2*a^6+3*a^5+3*a^4+a^3)*x3^2+(-2*B*C*a^7-a^5-2*a^4-2*a^3-a^2)*x3-C^2*a^7+
+a^3+a^2+a
```

The conditions under which c2 is null can be found with the following code:

```
c1=(-2*A*a^7-a^6-3a^7);
A=solve('c1','A');
c2=(a^5+4*a^6+3*a^7-a^7*(A^2+2*B));
B=solve('c2','B');
c3=(-a^7*(2*C+2*A*B)-a^4-3*a^5-5*a^6-a^7);
C=solve('c3','C');
c4=((-B^2-2*A*C)*a^7+2*a^6+3*a^5+3*a^4+a^3);
a1=solve('c4','a');
c5=(-2*B*C*a^7-a^5-2*a^4-2*a^3-a^2);
a2=solve('c5','a');
c6=-C^2*a^7+a^3+a^2+a;
a3=solve('c6','a');
```

The three vectors `a1`, `a2`, and `a3` contain the solutions with respect to `a` of the last three coefficients of polynomial `rem2`. Among them, we can find the unique common solution, that is $a = 1 + \sqrt{8}$.

2.5   Bifurcation diagram and Feigenbaum constant

The previous paragraphs have highlighted how the behavior of the logistic map strongly depends on the parameter a. This dependence can be system-

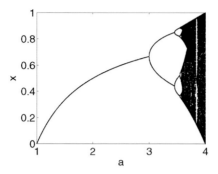

FIGURE 2.15
Bifurcation diagram of the logistic map for $a \in [1, 4]$.

atically explored by varying at small steps the parameter and observing the steady-state behavior obtained. In this way a bifurcation diagram of the system is constructed. The bifurcation diagram, thus, shows how the dynamical behavior of a system changes when a control parameter is varied. Numerical methods are often used to generate bifurcation diagrams. The bifurcation diagram of the logistic map is built in the next example.

Example 2.10 _____

By using MATLAB®, derive the bifurcation diagram for the logistic map.
Solution. To obtain the bifurcation diagram, it suffices to iterate the map for increasing values of the parameter a, each time starting again from a given initial condition. The following code may be used for the purpose:

```
n=1000;
x=zeros(n+1,1);
x(1)=0.51;
for a=1:0.002:4
for i=1:n
x(i+1)=a*x(i)*(1-x(i));
end
plot(a,x(900:end),'k.','MarkerSize',4)
hold on
end
```

The bifurcation diagram is shown in Figure 2.15.

The bifurcation diagram provides a comprehensive picture of the overall behavior of the system. Details can be further explored by zooming into the bifurcation diagram in the region of interest of the parameter, as shown in the next example.

Example 2.11 _____

By changing the interval of values in the procedure presented in the previous example, a magnification of the bifurcation diagram can be obtained. This is shown in Figure 2.16, where $a \in [3, 4]$. The bifurcation diagram reveals a sequence of period doublings that ultimately leads the system to a chaotic behavior.

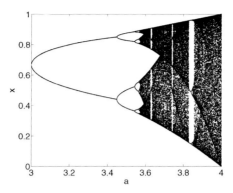

FIGURE 2.16
Bifurcation diagram of the logistic map for $a \in [3, 4]$, showing several windows of different periodicity.

By studying in detail the phenomenon of period doubling discussed above, Feigenbaum has discovered a regularity which is illustrated by further magnifying the bifurcation diagram in the region of the parameter $a \in [2.9, 3.6]$ as shown in Figure 2.17.
As we are interested in monitoring the sequence of bifurcations, let a_2, a_4, a_8, a_{16} indicate the bifurcation points ($a_2 = 3$ is the critical value of the parameter where the system bifurcates from stable equilibrium to period-2 cycle, $a_4 \approx 3.449$ the point where the transition from period-2 to period-4 cycle occurs, $a_8 \approx 3.544$ the parameter value at the bifurcation between period-4 and period-8 cycle, and so on). Let δ_k indicate the difference between two successive bifurcation points, that is, $\delta_1 = a_4 - a_2$, $\delta_2 = a_8 - a_4$ and so on. Having defined these parameters, one may notice the regularity in the sequence of bifurcations, a discovery due to Feigenbaum. In particular, considering the ratio $\frac{\delta_{k-1}}{\delta_k}$, the limit

$$\delta = \lim_{k \to \infty} \frac{\delta_{k-1}}{\delta_k}$$

defines the number indicated as δ and known as the Feigenbaum constant. At first approximation the Feigenbaum constant δ can be evaluated as:

$$\delta = \frac{a_4 - a_2}{a_8 - a_4} = \frac{3.449 - 3}{3.544 - 3.449} = 4.7263 \to_{n \to \infty} 4.6692 \qquad (2.24)$$

or considering the bifurcation points of the higher-order period cycles as:

$$\delta = \frac{a_4 - a_2}{a_8 - a_4} = \frac{3.544 - 3.449}{3.565 - 3.544} = 4.5238 \to_{n \to \infty} 4.6692 \qquad (2.25)$$

The Feigenbaum constant appears in many different systems and is a universal feature of chaos. It represents a fascinating property of a very common route from order to chaos: the period doubling route to chaos.

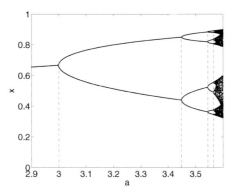

FIGURE 2.17
Bifurcation diagram of the logistic map for $a \in [2.9, 3.6]$, illustrating the phenomenon of period doubling.

2.6 Characterizing elements of chaotic behavior

Consider the continuous-time system

$$\dot{\mathbf{x}} = f(\mathbf{x}, t) \qquad (2.26)$$

with $\mathbf{x} \in \mathbb{R}^n$ and the discrete-time system:

$$\mathbf{x}_{k+1} = h(\mathbf{x}_k, k) \qquad (2.27)$$

with $\mathbf{x}_k \in \mathbb{R}^n$.

These dynamical systems are nonlinear and, as previously shown for the logistic map, they can exhibit chaotic motion. This particular regime is characterized by:

- aperiodic oscillations of the state variables;

- high sensitivity to initial conditions;

- sensitivity to parameter changes;

- period doubling cascades, if any, ruled by the Feigenbaum constant;

- long-term unpredictability;

- signals with a wide spectrum, similar to that of white noise.

These items represent qualitative features that may appear in chaotic systems. For a quantitative characterization of chaos, appropriate measures are introduced. The most important is the spectrum of Lyapunov exponents. We will introduce it referring to the logistic map, which is one-dimensional and discrete-time, but Lyapunov exponents can be defined and computed in the general case of multivariable and continuous-time systems.

2.6.1 Lyapunov exponents

The Lyapunov exponents are a measure quantifying the high sensitivity of chaotic systems to initial conditions. Recall Example 2.6; in that case the logistic map was started from two initial conditions differing for a small quantity $\varepsilon = 10^{-5}$, but the resulting trajectories were very different. Lyapunov exponents aim at measuring the rate of divergence of nearby trajectories. They are in a number equal to that of the state variables, so in the case of the logistic map one Lyapunov exponent, indicated as λ, is calculated.

Let x_i indicate the solution obtained for initial condition $x(0)$ and let x_i' the one starting from $x(0) + \varepsilon$. Let e_i indicate the difference between the two nearby trajectories, that is $e_i = x_i' - x_i$. Suppose that the difference e_{i-1} is small, the next sample is calculated by considering only first-order terms in the expansion of $f(x)$, that is

$$e_i = f'(x_{i-1})e_{i-1} \tag{2.28}$$

We now want to quantify how the difference in the initial conditions, e_0, propagates at time n. To do this, we consider $|e_n/e_0|$ and, taking into account (2.28), we write it as:

$$\left| \frac{e_n}{e_0} \right| = \left| \frac{e_n}{e_{n-1}} \right| \left| \frac{e_{n-1}}{e_{n-2}} \right| \cdots \left| \frac{e_1}{e_0} \right| \tag{2.29}$$

As we expect that $|e_n/e_0|$ grows exponentially, that is $|e_n/e_0| \simeq e^{\lambda n}$, we take the natural logarithm of Equation (2.29):

$$\ln \left| \frac{e_n}{e_0} \right| = \sum_{k=1}^{n} \ln \left| \frac{e_k}{e_{k-1}} \right| \tag{2.30}$$

Finally, the Lyapunov exponent $\lambda = \lambda(x_0)$ is defined as:

$$\lambda(x_0) = \lim_{n \to \infty} \frac{1}{n} \ln \left| \frac{e_n}{e_0} \right| = \lim_{n \to \infty} \frac{1}{n} \sum_{k=1}^{n} \ln \left| \frac{e_k}{e_{k-1}} \right| \tag{2.31}$$

In virtue of (2.28) it is equal to:

$$\lambda(x_0) = \lim_{n \to \infty} \frac{1}{n} \sum_{k=1}^{n} \ln |f'(x_{k-1})| \tag{2.32}$$

For the logistic map one has:

$$\lambda(x_0) = \lim_{n\to\infty} \frac{1}{n} \sum_{k=1}^{n} \ln|a - 2ax_{k-1}| = \ln a + \lim_{n\to\infty} \frac{1}{n} \sum_{k=1}^{n} \ln|1 - 2x_{k-1}| \quad (2.33)$$

Consider now the logistic map with $a = 4$. In this case, for the two fixed points $x(0) = 0$ or $x(0) = 1$ one gets $\lambda(0) = \lambda(1) = 1.39$. Furthermore, for all initial conditions $x(0)$ in $0 < x(0) < 1$ leading to a trajectory ending in one of these fixed points one also gets $\lambda = 1.39$. Such initial conditions constitute a dense subset of the interval $[0, 1]$. With the exclusion of the two fixed points and this dense set, for almost all initial conditions $x(0)$ in $[0, 1]$ we get $\lambda(x(0)) = 0.693$. This means that if an initial point is taken at random in the interval $[0, 1]$, then the orbit (almost surely) covers densely the interval and the Lyapunov exponent is $\lambda(x(0)) = 0.693$.

The sign of the Lyapunov exponent is particularly important. In fact, $\lambda > 0$ indicates that an infinitesimally small initial difference is amplified during time. Therefore, as the value obtained for $a = 4$, that is, $\lambda(x_0) = \ln 2 \approx 0.693$, is positive, for this value of the parameter the logistic map is chaotic.

In the next exercise the Lyapunov exponent is calculated in the whole range of interest of the parameter a.

Exercise 2.3 _____

Compute the Lyapunov exponent of the logistic map as a function of the parameter a for $0 < x(0) < 1$ by using MATLAB®.
Solution. The Lyapunov exponent can be calculated adopting the following code:

```
n=1000;
a_vector=1:0.002:4;
lamba_a=zeros(length(a_vector),1);
for j=1:length(a_vector)
x=.1;
lyaptmp=0;
a=a_vector(j)
for i=1:n
y(i)=a*x*(1-x);
x=y(i);
lyap=(1/n)*log(abs(a-2*a*x));
lyaptmp=lyap+lyaptmp;
end
lambda_a(j)=lyaptmp;
end
plot(a_vector,lambda_a)
```

In Figure 2.18 the Lyapunov exponent $\lambda(a)$ with $a \in [1, 4]$ is represented.

2.7   Exercises

1. Consider the Tent map

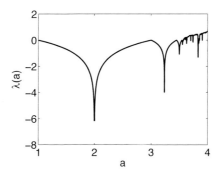

FIGURE 2.18
Lyapunov exponent of the logistic map for $a \in [1, 4]$ for $x(0) = 0.1$.

$$x_{k+1} = \begin{cases} 2x_k & \text{if } 0 \le x_k \le \frac{1}{2} \\ 2(1 - x_k) & \text{if } \frac{1}{2} < x_k \le 1 \end{cases} \qquad (2.34)$$

(a) Draw the map nonlinearity.

(b) Build a simple MATLAB® program to calculate the time series generated by the map.

(c) Compare it with the logistic map.

(d) Generalize the map by substituting the factor 2 appearing in Equation (2.34) with a parameter a. Then build the bifurcation diagram of the map with respect to this parameter and compare with that of the logistic map.

(e) Calculate and plot the Lyapunov exponents with respect to a.

Note: the solution of the initial value problem associated to the Tent map is $x_k = \frac{1}{\pi} \cos^{-1}(\cos(2^k \pi x(0)))$.

2. Consider the logistic map with $a = 4$ and initial condition $x(0) = \frac{1}{3}$. Derive without using the computer the time series $x(1), x(2), \ldots, x(n), \ldots$ and comment on the result obtained.

3. The asymmetric Tent map is defined as

$$x_{k+1} = \begin{cases} ax_k & \text{if } 0 \le x_k \le \frac{1}{a} \\ \frac{a}{a-1}(1 - x_k) & \text{if } \frac{1}{a} < x_k \le 1 \end{cases} \qquad (2.35)$$

where a, b, and c are parameters with $a > 0$, $b > 1$ and $a + b > ab$.

(a) Implement it.

(b) Derive a time series for given values of a, b, and c.

(c) Draw the bifurcation diagrams with respect to the parameters of the map.

(d) Derive the Lyapunov exponents as function of each parameter.

4. Consider the delayed logistic map

$$x_{k+1} = rx_k(1 - x_{k-1}) \tag{2.36}$$

It represents a population at the $k+1$ generation that depends not only on the population at the k generation but also on that at the $k-1$ generation.

(a) Draw the bifurcation diagram with respect to r.

(b) Plot x_k vs. x_{k-1}.

(c) Derive a surface plot reporting x_{k+1} as a function of x_k and x_{k-1}.

(d) Discuss the results obtained.

5. The following map is said to be the Bernoulli map:

$$x_{k+1} = f(x_k) \tag{2.37}$$

where $f(x_k) = 2x_k \mod 1$. Draw the time series generated by this map.

6. Make a qualitative comparison of the results derived analyzing the behavior of the maps at the points 1)-5).

7. Consider the cubic map $x_{k+1} = ax_k - x_k^3$.

(a) Find the equilibrium points and study their stability.

(b) Draw the bifurcation diagram with respect to a.

8. Consider two coupled logistic maps

$$\begin{aligned} x_{k+1} &= rx_k(1 - x_k) + \sigma(y_k - x_k) \\ y_{k+1} &= ry_k(1 - y_k) + \sigma(x_k - y_k) \end{aligned} \tag{2.38}$$

Derive the bifurcation diagrams considering fixed the parameter σ and varying r and vice versa.

9. Consider a natural number n and the following map

$$\begin{aligned} x_{k+2} &= \left\lfloor \frac{y_k + n}{y_{k+1}} \right\rfloor x_{k+1} - x_k \\ y_{k+2} &= \left\lfloor \frac{y_k + n}{y_{k+1}} \right\rfloor y_{k+1} - y_k \end{aligned} \tag{2.39}$$

with initial conditions $x_0 = 0$, $y_0 = 1$, $x_1 = 1$, $y_1 = n$, where $[x]$ indicates the largest integer not greater than x. x_i/y_i with $i = 2, 3, \ldots$ represents the Farey sequence. Analyze the map.

Further reading

For additional information on the topics of the chapter, the following references may be consulted: [50], [62], [63], [70], [81], [84].

3

Bifurcations

CONTENTS

In the previous chapter the logistic map has been discussed and its bifurcation diagram has been recalled several times. The diagram was used to illustrate how the behavior of the logistic map changes when the parameter a is varied. That was the first example of a bifurcation diagram; this chapter deals with bifurcations and the diagrams used to illustrate how and when they occur.

Even if in Chapter 2 the subject was introduced, continuing to move along the directions of the Varela's graph of Chapter 1, here bifurcations of first-order continuous-time systems will be dealt with, illustrating several types of bifurcations and their importance in the field of dynamical systems. In fact, bifurcations provide models of transitions and instabilities as some control parameters are varied.

3.1   Introduction to bifurcations in dynamical systems

Bifurcations sign the qualitative variations in the system behavior that may manifest when one or more parameters are varied. Let us consider for example a damped pendulum. A bifurcation occurs when the damping parameter is changed from zero to a non-null value. In fact, when there is no friction, the behavior of the pendulum is characterized by an oscillation with constant

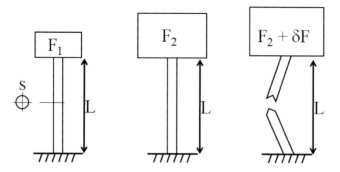

FIGURE 3.1
A structurally unstable column (large P).

amplitude. In the other case, friction causes a damping of the oscillation amplitude, a behavior that is clearly quite different from the previous one.

Bifurcation analysis detects the critical conditions when a qualitative shift of the system behavior appears. An important point in the analysis of the behavior of a system is to understand how and when a bifurcation occurs. Often, based on the physical knowledge of the process, from a primary investigation of the system it is straightforward to understand if some bifurcation occurs. For instance, bifurcations occur in structurally ill-conditioned systems. Let us consider, for example, the buckling of a column. The column has length L and section area S. Let us consider the parameter $P = L/S$. If P is large and a small weight is placed on top of the column, it can support the load and the column shape remains unaltered. Let us imagine increasing at small steps the load. At first the column will still be in vertical position. Then, for further increase the vertical position of the column will become unstable and a catastrophic behavior (the breakage of the column) will occur. Therefore, a thin column is a system that bifurcates (Figure 3.1). On the contrary, if P is small, the catastrophic behavior previously discussed does not occur. The column with small P is in fact structurally stable (Figure 3.2).

Finding the value of the critical force for the structurally unstable column is a classical problem of bifurcation, which was originally studied by Leonhard Euler (1707–1783) who was able to find the critical load P_E. Indicating with E the modulus of the elasticity of the column, with J the moment of inertia of the section area S, the critical load is:

$$P_E = \frac{\pi^2}{4} \frac{EJ}{L^2} \tag{3.1}$$

Below P_E the undeformed equilibrium position of the column is stable. It eventually loses its stability increasing the load slightly above P_E, when the bifurcation occurs.

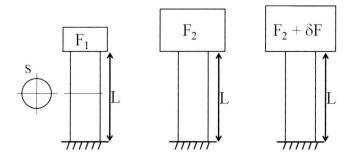

FIGURE 3.2
A structurally stable column (small P).

The first studies on bifurcations date back to Henri Poincaré (1854–1912) and are still an intense area of research. This chapter is dedicated to introducing the main notions for rather simple bifurcations. In particular, the so-called codimension-one bifurcations of equilibrium points will be dealt with. The term codimension-one refers to the fact that the occurrence of the bifurcation depends on a single parameter. Codimension-one bifurcations are introduced with first-order models, which are the simplest dynamical systems exhibiting such phenomena, but are relevant also for higher-order systems.

Example 3.1 _____

Let us discuss with some more detail the example related to the oscillations of a damped pendulum introduced above. The dynamical model of the damped pendulum can be written as:

$$\frac{d^2\theta}{dt^2} + \frac{\alpha}{ml}\frac{d\theta}{dt} + \frac{g}{ml}\sin(\theta) = 0 \tag{3.2}$$

where θ is the angular displacement of the pendulum with respect to the vertical axis, α is the damping coefficient, m is the mass, g is the gravity acceleration, and l is the length of the pendulum bar. The nonlinear model in Equation (3.2) can be approximated through a linear model by considering small values of angle θ, so that $\sin(\theta) \approx \theta$. The approximated model reads as:

$$\frac{d^2\theta}{dt^2} + \frac{\alpha}{ml}\frac{d\theta}{dt} + \frac{g}{ml}\theta = 0 \tag{3.3}$$

To simulate Equation (3.3) in MATLAB®, it has to be rewritten as a system of two first-order differential equations:

$$\begin{aligned} \dot{x} &= y \\ \dot{y} &= -\frac{\alpha}{ml}y + \frac{g}{ml}x \end{aligned} \tag{3.4}$$

where $\theta = x$ and $\dot{\theta} = y$. We first define the equations in the file `linpendulum.m` as:

```
function dxdt=linpendulum(t,x,alpha)
g=9.8;
l=1;
m=1
```

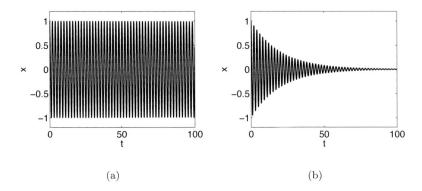

(a)   (b)

FIGURE 3.3
Bifurcation in a damped pendulum: (a) $\alpha = 0$ the pendulum oscillation maintains the same amplitude; (b) $\alpha = 0.1$ the pendulum oscillation damps towards the equilibrium.

```
dxdt=1*[x(2)
    -g/1/m*x(1)-alpha/1/m*x(2)];
```

and, then, use the routine **ode45** to integrate the dynamical system (3.4) passing to the function the value of the damping coefficient **alpha**

```
[T,Y]=ode45(@linpendulum,[0 100],[1 0],[],alpha);
```

The damping coefficient acts as a bifurcation parameter, in fact while $\alpha = 0$ leads to an undamped oscillation, as reported in Figure 3.3(a), varying the parameter to a non-null values, i.e., $\alpha = 0.1$, leads the system to an oscillation that is rapidly damped towards the equilibrium point $x = 0$, as shown in Figure 3.3(b).

3.2 Elementary bifurcations

Elementary bifurcations are studied assuming that the dynamical system admits an equilibrium point in $x = 0$ and representing the dynamics near the equilibrium as $\dot{x} = f(x, \mu)$, where μ is the bifurcation parameter defined so that the bifurcation occurs at $\mu = 0$. This kind of simplified representation takes the name of normal form. The normal form of a bifurcation is very important as it represents the elementary system showing the given bifurcation and, at the same time, the core mechanism in higher-order systems affected by the same bifurcation.

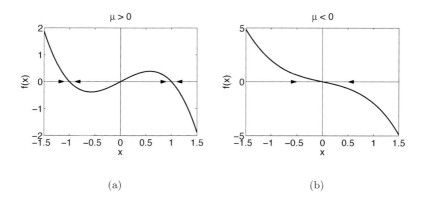

(a)   (b)

FIGURE 3.4
Graph of $f(x) = \mu x - x^3$: (a) $\mu = 1$; (b) $\mu = -1$.

3.2.1 Supercritical pitchfork bifurcation

The following differential equation characterizes the supercritical pitchfork bifurcation:

$$\dot{x} = \mu x - x^3 \tag{3.5}$$

where μ represents the bifurcation parameter. Let $f(x) = \mu x - x^3$. This function is plotted for $\mu < 0$ and for $\mu > 0$. MATLAB® commands are reported for two specific values of the parameter, i.e., $\mu = 1$ and $\mu = -1$:

```
mu=1;
x=[-1.5:0.01:1.5];
y=mu*x-x.^3;
figure,plot(x,y,'k','linewidth',2)
xlabel('x')
ylabel('f(x)')
title('\mu > 0')
hold on, plot([-1.5 1.5],[0 0],'k')
hold on, plot([0 0],[-2 2],'k')

mu=-1;
x=[-1.5:0.01:1.5];
y=mu*x-x.^3;
figure,plot(x,y,'k','linewidth',2)
xlabel('x')
ylabel('f(x)')
title('\mu < 0')
hold on, plot([-1.5 1.5],[0 0],'k')
hold on, plot([0 0],[-5 5],'k')
```

If $\mu > 0$ (Figure 3.4(a)), there are two equilibrium points, $\bar{x} = \pm\sqrt{\mu}$, that are stable (this can be shown by applying the graphical analysis of stability or noticing that $f'(x)|_{\bar{x}=\pm\sqrt{\mu}} = \mu - 3x^2|_{\bar{x}=\pm\sqrt{\mu}} = -2\mu$), and one unstable equilibrium $\bar{x} = 0$. Instead, if $\mu < 0$ (Figure 3.4(b)), there is only a (stable) equilibrium point $\bar{x} = 0$. $\mu = 0$ is the bifurcation point.

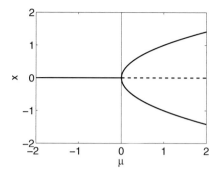

FIGURE 3.5
Diagram of the supercritical pitchfork bifurcation.

Let us now consider continuous changes of the parameter. At each value of μ we have to calculate the equilibrium points and their stability properties. In the case of the supercritical pitchfork bifurcation this can be done analytically as we have shown. Hence, the bifurcation diagram can be traced on the basis of the analytical results. We can use the following commands:

```
mu=[-2:0.01:0];
x=zeros(length(mu),1);
figure,plot(mu,x,'k','linewidth',2')
hold on
mu=[0:0.01:2];
x=zeros(length(mu),1);
plot(mu,x,'k--','linewidth',2')
x1=sqrt(mu);
x2=-sqrt(mu);
plot(mu,x1,'k','linewidth',2')
plot(mu,x2,'k','linewidth',2')
xlabel('\mu')
ylabel('x')
hold on, plot([0 0],[-2 2],'k')
```

to obtain the bifurcation diagram shown in Figure 3.5, where continuous lines denote stable branches, while dashed lines denote unstable branches. We observe that for $\mu < 0$ there is only one stable solution and for $\mu > 0$ two stable equilibrium points and one unstable. The number of equilibrium points changes at the critical value of the bifurcation $\mu = 0$. At this point the origin loses stability and becomes unstable, and at the same time two new (stable) equilibrium points are created ($\bar{x} = \pm\sqrt{\mu}$).

The bifurcation diagram can also be obtained numerically. This method is fundamental when analytical solutions are not available. So, let us now vary the parameter μ at small steps, starting from $\mu = -2$ to $\mu = 2$, and, for each value of the parameter, numerically calculate the solutions of $f(x)$ and their stability property through the evaluation of the Jacobian. To this aim, we can use the following MATLAB® commands:

```
for mu=-2:.001:2;
```

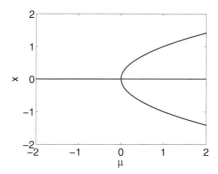

FIGURE 3.6
Numerical bifurcation diagram of the supercritical pitchfork bifurcation.

```
C=[-1 0 mu 0];
X=roots(C);
for i=1:3
    XI=imag(X(i,1));
    if XI==0
        JA=-3*X(i,1)^2+mu;
        if JA <0
        plot(mu,X(i,1),'.b');
        hold on;
        else plot(mu,X(i,1),'.r');
        end
    end
end
end
ylabel('x'), xlabel('\mu')
```

In the MATLAB® commands reported above, the bifurcation parameter has been changed in the interval $[-2, 2]$. For each value of μ the equilibrium points are calculated by solving $-x^3 + \mu x = 0$. Then, one has to check those that are real to plot them. To study the stability of each equilibrium point $\bar{x}$, the Jacobian is computed, that is $\text{JA} = f'(x)|_{x=\bar{x}} = -3\bar{x}^2 + \mu$. If JA is negative, then the equilibrium point is stable and is plotted in blue, otherwise it is plotted in red. The obtained bifurcation diagram is shown in Figure 3.6.

3.2.2 Subcritical pitchfork bifurcation

The following differential equation defines the subcritical pitchfork bifurcation:

$$\dot{x} = \mu x + x^3 \tag{3.6}$$

As for the supercritical case studied in Section 3.2.1, we can first inspect the number and stability properties of the equilibrium points by studying $f(x) = \mu x + x^3 = 0$. This gives for $\mu < 0$, three solutions and, for $\mu > 0$, one solution. As an example, we can consider $\mu = -1$ and $\mu = 1$ and display $f(x)$ in the two cases with the following MATLAB® commands:

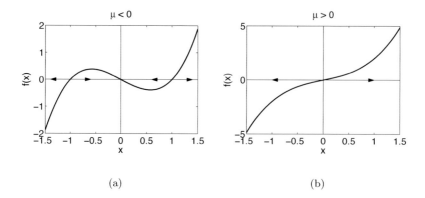

(a)   (b)

FIGURE 3.7
Graph of $f(x) = \mu x + x^3$: (a) $\mu = -1$; (b) $\mu = 1$.

```
mu=-1;
x=[-1.5:0.01:1.5];
y=mu*x+x.^3;
figure,plot(x,y,'k','linewidth',2)
xlabel('x'), ylabel('f(x)'), title('\mu < 0')
hold on, plot([-1.5 1.5],[0 0],'k')
hold on, plot([0 0],[-2 2],'k')

mu=1;
x=[-1.5:0.01:1.5];
y=mu*x+x.^3;
figure,plot(x,y,'k','linewidth',2)
xlabel('x'), ylabel('f(x)'), title('\mu > 0')
hold on, plot([-1.5 1.5],[0 0],'k')
hold on, plot([0 0],[-5 5],'k')
```

The equilibrium points are similar to those found for the supercritical pitchfork bifurcation, but with an important difference. As shown in Figure 3.7, now, for $\mu < 0$ three solutions appear: the two solutions $\bar{x} = \pm\sqrt{\mu}$ are now unstable, while $\bar{x} = 0$ is stable. For $\mu > 0$ there is only one unstable equilibrium point, that is $\bar{x} = 0$. The overall picture of the system behavior is clarified by the bifurcation diagram shown in Figure 3.8 and obtained with the MATLAB® commands:

```
mu=[-2:0.01:0];
x=zeros(length(mu),1);
figure,plot(mu,x,'k','linewidth',2')
hold on
x1=sqrt(-mu);
x2=-sqrt(-mu);
plot(mu,x1,'k--','linewidth',2')
plot(mu,x2,'k--','linewidth',2')
mu=[0:0.01:2];
x=zeros(length(mu),1);
plot(mu,x,'k--','linewidth',2')
xlabel('\mu'), ylabel('x')
hold on, plot([0 0],[-2 2],'k')
```

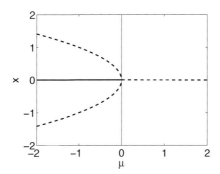

FIGURE 3.8
Diagram of the subcritical pitchfork bifurcation.

The shape of the diagram explains the origin of the name subcritical pitchfork bifurcation. The bifurcation is called subcritical as the pitchfork appears below the critical value of μ. Similarly, the bifurcation in Section 3.2.1 is called supercritical, as the pitchfork occurs above the critical value of μ.

Note that the subcritical bifurcation is obtained by changing the sign of a single term in the differential equation characterizing the supercritical bifurcation. This sign change yields substantial differences in the stability properties of the system (observe that the diagram of the subcritical pitchfork bifurcation can be obtained by rotating 180° that of the supercritical one and reversing the stability properties). While in the supercritical bifurcation there is always at least one stable equilibrium point, in the subcritical case there is a region of the parameter space where there are no stable equilibrium points. So, while in the supercritical case the bifurcation leads the system from one stable point to another, in the subcritical case the system may be led from a configuration where there is one stable equilibrium to a configuration where it cannot settle in any equilibrium point.

3.2.3 Saddle-node bifurcation

This type of bifurcation is characterized by a mechanism that leads to create, collapse, and annihilate equilibrium points. To describe this bifurcation we consider the following normal form:

$$\dot{x} = \mu - x^2 \tag{3.7}$$

We start the discussion by defining $f(x) = \mu - x^2$ and calculating the equilibrium points and their stability with the graphical analysis for three values of μ, one positive, one negative, and one zero:

```
mu=-1;
x=[-1.5:0.01:1.5];
```

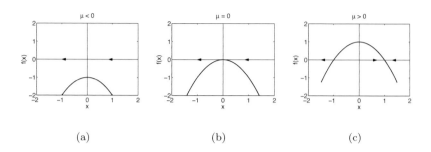

(a)   (b)   (c)

FIGURE 3.9
Graph of $f(x) = \mu - x^2$: (a) $\mu = -1$; (b) $\mu = 0$; (c) $\mu = 1$.

```
y=mu-x.^2;
figure,plot(x,y,'k','linewidth',2)
xlabel('x'), ylabel('f(x)'), title('\mu < 0')
hold on, plot([-2 2],[0 0],'k')
hold on, plot([0 0],[-2 2],'k')

mu=0;
x=[-1.5:0.01:1.5];
y=mu-x.^2;
figure,plot(x,y,'k','linewidth',2)
xlabel('x'), ylabel('f(x)'), title('\mu = 0')
hold on, plot([-2 2],[0 0],'k')
hold on, plot([0 0],[-2 2],'k')

mu=1;
x=[-1.5:0.01:1.5];
y=mu-x.^2;
figure,plot(x,y,'k','linewidth',2)
xlabel('x'), ylabel('f(x)'), title('\mu > 0')
hold on, plot([-2 2],[0 0],'k')
hold on, plot([0 0],[-2 2],'k')
```

The result is shown in Figure 3.9. We observe that for $\mu = 0$ there is a unique equilibrium point that is not stable. Perturbations on the right of the point, $x(0) \in \mathbb{R}^+$, push the system towards the equilibrium, while perturbation on the left, $x(0) \in \mathbb{R}^-$, lead away from the equilibrium.

For $\mu > 0$ there are two equilibrium points, one is stable and the other unstable. For $\mu < 0$, $f(x) = 0$ does not admit any solution, so there are no equilibrium points. This behavior can be visualized with the help of the bifurcation diagram, shown in Figure 3.10, obtained with the following commands:

```
mu=[0:0.01:2];
x1=sqrt(mu);
x2=-sqrt(mu);
figure,plot(mu,x1,'k','linewidth',2')
hold on,plot(mu,x2,'k--','linewidth',2')
xlabel('\mu'), ylabel('x')
plot([-2 2],[0 0],'k')
plot([0 0],[-2 2],'k')
```

Starting from negative values of μ and increasing it, we observe that at

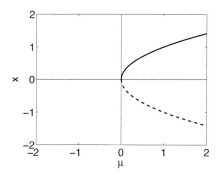

FIGURE 3.10
Saddle-node bifurcation diagram.

first the system has no equilibrium points, then as soon as the critical value $\mu = 0$ is crossed, two equilibrium points with opposite stability properties appear. These are $\bar{x} = \pm\sqrt{\mu}$.

The name given to this bifurcation derives from the nature of the equilibrium point at the critical value of the parameter. For $\mu = 0$, the point is not hyperbolic (in fact, both $f(x)$ and $f'(x)$ become zero for $x = 0$) and is a saddle-node fixed point. Given the form of the bifurcation diagram the saddle-node bifurcation is also known as fold bifurcation.

3.2.4 Transcritical bifurcation

In the transcritical bifurcation the equilibrium points (they are two) always persist, but when the bifurcation point is crossed they exchange their stability properties. The normal form showing this kind of bifurcation is:

$$\dot{x} = \mu x - x^2 \tag{3.8}$$

Consider now $f(x) = \mu x - x^2$. This function is plotted in Figure 3.11 based on the following MATLAB® commands:

```
mu=-1;
x=[-2:0.01:2];
y=mu*x-x.^2;
figure,plot(x,y,'k','linewidth',2)
xlabel('x'), ylabel('f(x)'), title('\mu < 0')
hold on, plot([-2 2],[0 0],'k')
hold on, plot([0 0],[-2 2],'k')

mu=0;
x=[-2:0.01:2];
y=mu*x-x.^2;
figure,plot(x,y,'k','linewidth',2)
xlabel('x'), ylabel('f(x)'), title('\mu = 0')
hold on, plot([-2 2],[0 0],'k')
hold on, plot([0 0],[-2 2],'k')
```

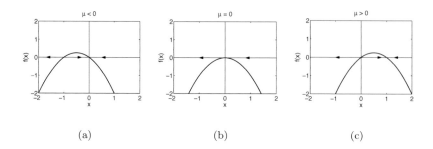

FIGURE 3.11
Graph of $f(x) = \mu x - x^2$: (a) $\mu = -1$; (b) $\mu = 0$; (c) $\mu = 1$.

```
mu=1;
x=[-2:0.01:2];
y=mu*x-x.^2;
figure,plot(x,y,'k','linewidth',2)
xlabel('x'), ylabel('f(x)'), title('\mu > 0')
hold on, plot([-2 2],[0 0],'k')
hold on, plot([0 0],[-2 2],'k')
```

We report here the commands for the bifurcation diagram (shown in Figure 3.12).

```
mu=[-2:0.01:0];
x1=zeros(length(mu),1);
x2=mu;
figure,plot(mu,x1,'k','linewidth',2')
hold on,plot(mu,x2,'k--','linewidth',2')
mu=[0:0.01:2];
x1=zeros(length(mu),1);
x2=mu;
plot(mu,x1,'k--','linewidth',2')
plot(mu,x2,'k','linewidth',2')
xlabel('\mu'), ylabel('x')
plot([0 0],[-2 2],'k')
```

From the analysis of the function $f(x)$, its equilibrium points and their stability and from the bifurcation diagram, we note that for any μ two equilibrium points exist: $\bar{x} = 0$ and $\bar{x} = \mu$ (they coincide at the critical value of the parameter, that is $\mu = 0$). When $\mu = 0$ is crossed their properties of stability change and, in particular, $\bar{x} = 0$ is stable for $\mu < 0$ and unstable for $\mu > 0$, while $\bar{x} = \mu$ is unstable for $\mu < 0$ and stable for $\mu > 0$.

3.2.5   Perturbed subcritical pitchfork bifurcation

We have seen that the subcritical pitchfork bifurcation is characterized by the system $\dot{x} = \mu x + x^3$. In this system for $\mu > 0$ the unique equilibrium point ($\bar{x} = 0$) is unstable. We study here the effect of a further, higher-order polynomial term in the system dynamics. In particular, we consider

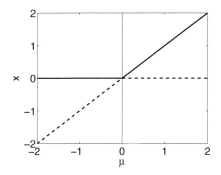

FIGURE 3.12

Diagram of the transcritical bifurcation.

$$\dot{x} = \mu x + x^3 - x^5 \tag{3.9}$$

The bifurcation appearing in system (3.9) is called the perturbed subcritical pitchfork bifurcation. The bifurcation diagram is shown in Figure 3.13: stable equilibrium points are drawn in blue, while unstable ones are in red. To derive it the following MATLAB® commands have been used:

```
for mu=-1:.001:1;
    C=[-1 0 1 0 mu 0];
    X=roots(C);
    for i=1:5
        XI=imag(X(i,1));
        if XI==0
            JA=-5*X(i,1)^4+3*X(i,1)^2+mu;
            if JA <0
            plot(mu,X(i,1),'.b');
            hold on;
            else plot(mu,X(i,1),'.r');
            end
        end
    end
end
ylabel('x')
xlabel('\mu')
```

The bifurcation diagram allows us to draw the following considerations. The effect of the perturbation is stabilizing; in fact, for $\mu > 0$ two stable equilibrium points are obtained. The system behavior has a hysteresis; in fact, increasing μ from negative values the origin remains stable until $\mu = 0$. At this point, the equilibrium loses stability and the system jumps into one of the other two equilibrium points. If μ is now decreased, the system moves into a region where there are three stable equilibrium points; it will remain into the upper or lower branch until the two outer equilibrium points disappear. Only at this point will the system jump back to the origin.

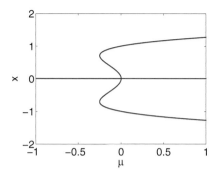

FIGURE 3.13
Diagram of the perturbed subcritical pitchfork bifurcation.

3.2.6   Imperfect bifurcations

In Section 3.2.5 the stabilizing effect of odd polynomial uncertainty in the subcritical pitchfork bifurcation has been discussed. Let us consider what happens in the case of supercritical pitchfork bifurcation when a further additive constant term is included. This is an example of imperfect bifurcation, where the system behavior is analyzed with respect to two parameters. The supercritical pitchfork bifurcation with the further term reads

$$\dot{x} = r + \mu x - x^3 \tag{3.10}$$

where r and μ are the parameters.

To investigate the equilibrium point of system (3.10) we have to consider

$$r + \mu x - x^3 = 0 \tag{3.11}$$

that is equivalent to solving the system

$$\begin{aligned} y &= \mu x - x^3 \\ y &= -r \end{aligned} \tag{3.12}$$

that clearly shows that the equilibrium points are obtained as the intersection of two curves $y = \mu x - x^3$ and $y = -r$. Note that the parameter r breaks the symmetry of the perfect supercritical pitchfork bifurcation.

Let us discuss the case $\mu \leq 0$. In this case, for any value of r the solution is unique. In addition we note that the equilibrium point is stable. An example is shown in Figure 3.14, obtained with the following commands (with $\mu = -1$ and $r = -2$):

```
mu=-1;
r=2;
x=[-1.5:0.01:1.5];
y=mu*x-x.^3;
```

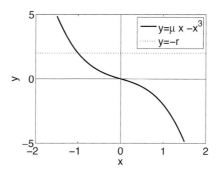

FIGURE 3.14

Equilibrium point for system (3.10) with $\mu = -1$ and $r = -2$.

```
figure,plot(x,y,'k','linewidth',2)
hold on, plot([-2.5 2.5],[r r],'k:','linewidth',2)
xlabel('x'), ylabel('y')
hold on, plot([-2.5 2.5],[0 0],'k')
hold on, plot([0 0],[-5 5],'k')
```

The most interesting scenario occurs when $r > 0$ as shown in the next example.

Example 3.2 _____

Plot the curves $y = \mu x - x^3$ with $\mu = 1$ and $y = -r$ for different values of r.
Solution. Consider Figure 3.15, drawn with the following commands:

```
mu=1;
x=[-1.5:0.01:1.5];
y=mu*x-x.^3;
figure,plot(x,y,'k','linewidth',2)
r=1;
hold on, plot([-1.5 1.5],[r r],'k:','linewidth',2)
r=2/(3*sqrt(3));
hold on, plot([-1.5 1.5],[r r],'k--','linewidth',2)
r=0.1;
hold on, plot([-1.5 1.5],[r r],'k-.','linewidth',2)
xlabel('x'), ylabel('y')
hold on, plot([-1.5 1.5],[0 0],'k')
hold on, plot([0 0],[-2 2],'k')
```

We note that there exist values of r such that a unique (stable) equilibrium point is found; for other values of r there are three equilibrium points (two stable and one unstable), and, finally, there are two critical values of r ($r = \pm\frac{2}{3\sqrt{3}}$) for which two equilibrium points are found (in this case the stable and unstable points collide into a saddle-node).

To obtain analytically the critical value of the parameter r, say r_c, in the more general case, one has to first find the local minima/maxima x_m of the curve $y = \mu x - x^3$:

$$\frac{d}{dx}(\mu x - x^3) = 0 \Rightarrow x_m = \pm\sqrt{\frac{\mu}{3}} \tag{3.13}$$

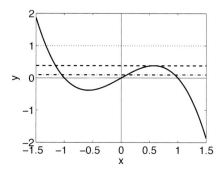

FIGURE 3.15
Equilibrium point for system (3.10) with $\mu = 1$ and different values of r: $r = 1$ (dotted line), $r = \frac{2}{3\sqrt{3}}$ (dashed line), and $r = 0.1$ (dash-dotted line).

and thus

$$y_m = \pm \frac{2\mu}{3}\sqrt{\frac{\mu}{3}} \qquad (3.14)$$

Hence, the critical value of the parameter is

$$r_c = \pm \frac{2\mu}{3}\sqrt{\frac{\mu}{3}} \qquad (3.15)$$

The number of equilibrium points for system (3.10) in the parameter space $\mu - r$ is summarized in Figure 3.16, obtained with the MATLAB® commands:

```
axis([-5 5 -5 5]);
hold on;
mu=[.0001:.001:5];
rc=(2/3)*mu.*(mu./3).^(.5);
plot(mu,rc,'k','linewidth',2);
hold on;
plot(mu,-rc,'k','linewidth',2);
ylabel('r'), xlabel('\mu')
```

Each region of the parameter space indicates the number of equilibrium points found in system (3.10) for the values of the parameters corresponding to the abscissa and ordinate. On the locus the equilibrium points are two. For $\mu = r$ there is a cusp point. Even if in the other points of the locus a saddle-node bifurcation occurs, only in the cusp point (achieved while varying at the same time both μ and r) a single stable equilibrium point is obtained. To get further insight into this property of the imperfect supercritical pitchfork bifurcation, we derive a set of bifurcation diagrams with respect to μ and for several values of r.

Consider the following commands:

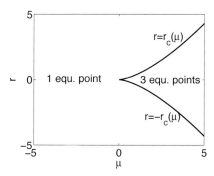

FIGURE 3.16
Number of equilibrium points for system (3.10) in the parameter space $\mu - r$.

```
figure
hold on;
for r=.1:.5:2.5;
for mu=-5:.01:5;
    C=[-1 0 mu r];
    X=roots(C);
    for i=1:3;
        XI=imag(X(i,1));
        if XI==0;
            JA=-3*X(i,1)^2+mu;
            if JA <0
                plot(mu,X(i,1),'ob','MarkerSize',2);
                else plot(mu,X(i,1),'or','MarkerSize',2); end
        end
    end
end
end
xlabel('\mu'), ylabel('x')
```

The result is reported in Figure 3.17, where blue curves indicate stable branches, while red curves show unstable branches.

3.3 Bifurcations towards catastrophes

As has been previously discussed, the bifurcation theory studies systems characterized by equations that qualitatively change their behavior depending on one or more parameters. In particular, it has been shown that, when in the space of parameters some singularity like a cusp occurs, even a small change in the parameters may lead to the appearance or disappearance of equilibrium points or to a change of the nature of the equilibrium point, from attractor to repellor, for instance. These scenarios, generally characterized by well-defined geometrical structures, lead to the catastrophes. The theory was introduced

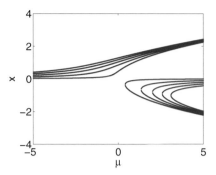

FIGURE 3.17
Bifurcation diagrams of system (3.10) with respect to μ for several values of r.

by René Thom in the 1960s and great contributions to the topic were given by Christopher Zeemen in the 1970s [95]. We will discuss it with an outstanding example reported by Steven Strogatz in his book [84], that is well conceived to introduce the reader to the problem.

The example refers to the modeling of the sudden outbreak of an insect population that leads to defoliation and destruction of forests of fir trees. Typical time-scales of this problem are related to the time needed by trees to change all the leaves, which is in the order of 7–10 years, and to the time needed by insects to destroy a forest, an event that might occur in 4 years. Without insects, the forest may survive for 100–150 years.

The size N of the insect population can be modeled by the following equation:

$$\dot{N} = RN\left(1 - \frac{N}{K}\right) - p(N) \tag{3.16}$$

where the term $RN(1 - \frac{N}{K})$ represents the insect population growth without predation, while $p(N)$ is the predation effect. R is the growth rate of the insect population and K the carrying capacity of the biological species in the environment, i.e., the maximum population size that the environment can sustain indefinitely. The predation effect is modeled by

$$p(N) = \frac{BN^2}{A^2 + N^2} \tag{3.17}$$

This function saturates as both the number of predators and their foraging capacity are limited. A and B are positive constants.

The model comprises four parameters: R, K, A, and B. Properly scaling of the model through the introduction of the following quantities

$$x = \frac{N}{A}$$
$$\tau = \frac{Bt}{A}$$
$$r = \frac{RA}{B}$$
$$k = \frac{K}{A}$$
(3.18)

yields a system

$$\frac{dx}{d\tau} = rx\left(1 - \frac{x}{k}\right) - \frac{x^2}{1+x^2}$$
(3.19)

with two independent parameters r and k. To calculate the equilibrium points, $\frac{dx}{dt} = 0$ must hold. It follows that $x = 0$ is a trivial equilibrium point. The other equilibrium points are calculated by solving

$$r\left(1 - \frac{x}{k}\right) = \frac{x}{1+x^2}$$
(3.20)

or, equivalently, finding the intersections of the two curves

$$y = r(1 - \frac{x}{k})$$
$$y = \frac{x}{1+x^2}$$
(3.21)

The first curve is a straight line. The mutual position of this line and the curve $y = \frac{x}{1+x^2}$ depends on the parameters r and k and determines the number and values of the equilibrium points of the model (3.19).

Assuming both k and r small (for instance, $k = 0.4$ and $r = 1.5$) a unique stable equilibrium point is obtained as in Figure 3.18:

```
k=0.4;
r=1.5;
x=[0:0.1:100];
y=x./(1+x.^2);
y2=k*(1-x/r);
figure,plot(x,y,'k','linewidth',2)
hold on,plot(x,y2,'r','linewidth',2)
xlim([0 7]), ylim([0 0.6])
xlabel('x'), ylabel('y')
```

This equilibrium represents a stable, relatively small population.

Consider now $k = 0.5$ and $r = 12$:

```
k=0.5;
r=12;
x=[0:0.1:20];
y=x./(1+x.^2);
y2=k*(1-x/r);
figure,plot(x,y,'k','linewidth',2)
hold on,plot(x,y2,'r','linewidth',2)
xlim([0 7])
ylim([0 0.6])
xlabel('x')
ylabel('y')
```

Figure 3.19 shows that under these conditions three equilibrium points, labelled as a, b, and c, are obtained. The graphical analysis for studying their stability properties is illustrated in Figure 3.20 obtained with the following MATLAB® commands:

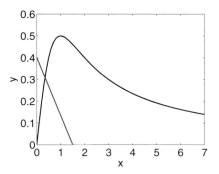

FIGURE 3.18
Equilibrium points for system (3.19) with $k = 0.4$ and $r = 1.5$.

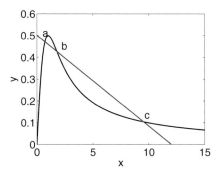

FIGURE 3.19
Equilibrium points for system (3.19) with $k = 0.5$ and $r = 12$.

```
k=0.5;
r=12;
x=[0:0.1:20];
y=k*x.*(1-x/r)-x.^2./(1+x.^2);
figure,plot(x,y,'k','linewidth',2)
xlim([0 15]), ylim([-0.6 0.6])
hold on, plot([0 15],[0 0],'k')
xlabel('x'), ylabel('f(x)')
```

The points a and c are stable, while b is unstable. The stability of point c is particularly relevant, as this represents a large outbreak of the insects that could potentially destroy the forest. Each initial condition between point b and c is catastrophic as starting from it the system will reach point c.

The bifurcation point where b and c collapse into a single equilibrium point occurs when the straight line $y = r(1 - \frac{x}{k})$ becomes tangent to the curve $y = \frac{x}{1+x^2}$. This is verified if the following conditions hold:

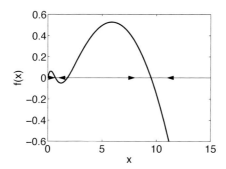

FIGURE 3.20
Stability graphical analysis for the equilibrium points for system (3.19) with $k = 0.5$ and $r = 12$.

$$r(1 - \tfrac{x}{k}) = \tfrac{x}{1+x^2}$$
$$r\tfrac{d}{dx}(1 - \tfrac{x}{k}) = \tfrac{d}{dx}\tfrac{x}{1+x^2} \qquad (3.22)$$

The first equations represent the intersection condition, the second one is the tangency condition. From the second of (3.22), we have

$$-\frac{r}{k} = \frac{1 - x^2}{(1 + x^2)^2} \qquad (3.23)$$

and substituting this into the first of (3.22), we get

$$r = \frac{2x^3}{(1 + x^2)^2} \qquad (3.24)$$

and, so, for the parameter k

$$k = \frac{2x^3}{x^2 - 1} \qquad (3.25)$$

By using the following commands

```
axis([0 40 0 .8]);
hold on;
xv=[0:0.001:20];
for i=1:length(xv);
    x=xv(i);
    r=2*x^3/(1+x^2)^2;
    k=2*x^3/(x^2-1);
    plot(k,r,'k.');
    hold on;
end
xlabel('k')
ylabel('r')
```

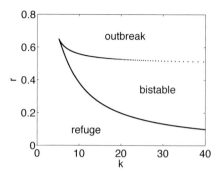

FIGURE 3.21
Bifurcation diagram for system (3.19).

the bifurcation diagram shown in Figure 3.21 is obtained. In the bifurcation diagram there is a cusp point, the fingerprint of catastrophes. The outbreak region is dangerous for the forest, since it would lead to a sensible and unsustainable increase of the insect population size, while the optimal scenario is represented by the region of parameters labeled as "refuge." In the bistability region the survival of the forest depends on the initial condition.

3.4 Exercises

1. Classify the following systems referring to the type of simple bifurcation they exhibit:

   (a) $\dot{x} = x(\mu - e^{\alpha x})$;

   (b) $\dot{x} = \mu + x + \frac{1}{\log(1+x)}$;

   (c) $\dot{x} = x + \tanh(\mu x)$;

   (d) $\dot{x} = \mu - x - e^{-x}$;

   (e) $\dot{x} = \mu x + x^3 - x^5 + x^7$;

   (f) $\dot{x} = \mu x - x^3 + x^5 - x^7$.

2. Search on internet the Ising model for magnetization and link it to bifurcation theory.

3. Consider the system

$$\dot{x} = -x + \mu \tanh x$$

   Numerically find the bifurcation diagram of the system and classify its type.

4. For the system:
$$\dot{x} = \mu x + x^5$$
identify the type of bifurcation and draw the corresponding diagram.

5. Consider the second-order system:
$$\ddot{x} + \mu\dot{x} - x + x^2 = 0$$
Discuss the possible bifurcations of the equilibrium points of the system.

6. Repeat the previous exercise for the system:
$$\ddot{x} + \dot{x} + \mu x + x^3 = 0$$

7. Consider the system in Figure 3.22 illustrating a bead constrained to move along a hoop, which in turn rotates around its vertical axis with angular velocity Ω. The bead is also subjected to gravitational force. The following system:

$$\ddot{\theta} + \mu\dot{\theta} + (\omega^2 - \Omega^2 \cos\theta)\sin\theta = 0$$

describes the motion of the particle in terms of the angle between the bead and the rotation axis. μ is the damping coefficient, $\omega^2 = \frac{g}{r}$, and g is the gravity.

(a) Determine the equilibrium points for $\mu = 0$.

(b) Analyze the system for $\mu \neq 0$ and derive the bifurcation diagram with respect to Ω.

8. Propose perturbed models of saddle-node bifurcation.

9. Consider a fork bifurcation. Assume that it represents a ball moving on a surface. Provide a description of the phenomenon in 3D.

10. Describe what is a cusp, taking into account the information presented in the chapter.

11. Consider the mapping of a point (x_1, x_2) of the Euclidean space into (y_1, y_2) according to the equations:

$$y_1 = x_1^2$$
$$y_2 = x_2$$

This mapping is referred as Whitney fold mapping. Write a MATLAB® procedure to draw this map.

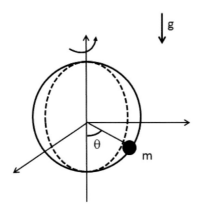

FIGURE 3.22
An overdamped bead on a rotating hoop.

12. Consider now the Whitney cusp mapping

$$y_1 = x_1^3 + x_1 x_2$$
$$y_2 = x_2$$

write a MATLAB® procedure to draw it.

Further reading

For additional information on the topics of the chapter, the following references may be consulted: [37], [49], [84], [86], [96].

4

Oscillators

CONTENTS

The elementary bifurcations discussed and analyzed in the previous chapter clarify that dramatic changes in nonlinear system behavior can be induced by slightly varying even a single parameter. Up to now we have dealt with first-order systems. In this Chapter we aim to move along the diagram introduced in Chapter 1 increasing the dimension of the system, i.e., the number of state variables. Second-order systems are considered unveiling their fundamental importance in the study of oscillations and oscillators.

4.1 Hopf bifurcation

Before entering in the core subject of this chapter, we introduce the Hopf bifurcation, which appears in systems of order equal or greater than two. This bifurcation links the previous chapter, dealing with elementary bifurcations, with this focused on oscillators. The Hopf bifurcation will be shortly discussed,

but the topic is so important that entire books (for example, [57]) are devoted
to illustrate it.

In Chapter 3 bifurcations of the equilibrium points in first-order systems
have been dealt with. Instead, discussing the Hopf bifurcation requires a system of order two:

$$\dot{x}_1 = \mu x_1 - x_2 - x_1(x_1^2 + x_2^2)$$
$$\dot{x}_2 = x_1 + \mu x_2 - x_2(x_1^2 + x_2^2) \tag{4.1}$$

This system represents the Hopf normal form and is also referred to as the
Hopf oscillator. The point $(x_1, x_2) = (0,0)$ is an equilibrium point for $\forall\ \mu$.
The Jacobian of system (4.1) is

$$J = \frac{\partial f}{\partial x} = \begin{pmatrix} \mu - 3x_1^2 - x_2^2 & -1 - 2x_1 x_2 \\ 1 - 2x_1 x_2 & \mu - x_1^2 - 3x_2^2 \end{pmatrix} \tag{4.2}$$

When calculated around the equilibrium point $(x_1, x_2) = (0,0)$, the Jacobian gives a pair of complex eigenvalues: $\lambda = \mu \pm j$. Therefore, the equilibrium
point is stable for $\mu < 0$ and unstable for $\mu > 0$. $\mu = 0$ is the bifurcation point
where a *limit cycle* is created. A limit cycle is a closed orbit in the phase
space, that is, a trajectory not constant, but returning to its starting point.

A limit cycle can be stable (if all neighboring trajectories approach it
as time goes to infinity), unstable (if, instead, all neighboring trajectories
approach it as time goes to negative infinity), or neither stable nor unstable.
The limit cycle obtained in system (4.1) for $\mu > 0$ is stable.

Stable limit cycles are fundamental in all sciences. In electronics, the design
of circuits having stable limit cycles is a fundamental topic. Stable limit cycles
maintain oscillations in a system.

System (4.1) may be rewritten in polar coordinates with the transforma-
tion

$$x_1 = \rho \cos \theta$$
$$x_2 = \rho \sin \theta \tag{4.3}$$

so that it reads

$$\dot{\rho} = \rho(\mu - \rho^2)$$
$$\dot{\theta} = 1 \tag{4.4}$$

Note that the first equation of (4.4) is that of a saddle-node bifurcation.
$\rho = \sqrt{\mu}$ is a solution as $\dot{\rho} = 0$, while $\dot{\theta} = 1$ is solved by $\theta = t$. This solution
represents an oscillation of the system. In the original system the solution
reads

$$x_1 = \sqrt{\mu} \cos t$$
$$x_2 = \sqrt{\mu} \sin t \tag{4.5}$$

which clearly shows that the oscillations have a constant amplitude $\rho = \sqrt{\mu}$
and phase $\theta = t$ (in the solution, without lack of generality, the initial phase
has been set to zero).

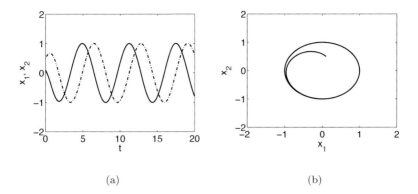

(a)   (b)

FIGURE 4.1
Numerical simulation of system (4.1) with $\mu = 1$. (a) Waveforms of the state variables. (b) Trajectory in the plane $x_1 - x_2$.

Example 4.1 _____

Numerically integrate system (4.1) and derive the bifurcation diagram.
Solution. The equations of system (4.1) are written in the file HopfOsc.m:

```
function dxdt=HopfOsc(t,x)

mu=1;
F=(mu-(x(1)^2+x(2)^2));

dxdt=[-x(2)+F*x(1);
x(1)+x(2)*F];

end
```

where, for the sake of illustration, the parameter μ was fixed to $\mu = 1$.
The command ode45 is used to integrate the system for $t \in [0, 20]$ and starting from initial conditions $(x_1, x_2) = (0.1, 0.5)$:

```
[t,y]=ode45(@HopfOsc,[0:0.001:20],[0.1, 0.5]);
```

After the integration, the waveforms of the state variables and the trajectory in the plane $x_1 - x_2$ are visualized with the commands:

```
figure,plot(t,y(:,1),'k','linewidth',2)
hold on, plot(t,y(:,2),'k-.','linewidth',2)
xlabel('t')
ylabel('x_1, x_2')
figure,plot(y(:,1),y(:,2),'k','linewidth',2)
xlabel('x_1')
ylabel('x_2')
```

The result is shown in Figure 4.1. Note that the amplitude of the limit cycle is $\sqrt{\mu} = 1$.
To obtain the bifurcation diagram shown in Figure 4.2, first we rewrite the function containing the system equations so that μ can be given as an external parameter (the function is renamed HopfOscMu):

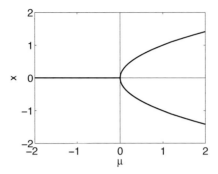

FIGURE 4.2
Bifurcation diagram of system (4.1).

```
function dxdt=HopfOscMu(t,x,mu)

F=(mu-(x(1)^2+x(2)^2));

dxdt=[-x(2)+F*x(1);
x(1)+x(2)*F];

end
```

then, the following MATLAB® commands are used:

```
mu=[-2:0.01:2];
xM=zeros(length(mu),1);
xm=zeros(length(mu),1);
for i=1:length(mu)
[t,y]=ode45(@HopfOscMu,[0:0.01:500],[0.01,  -0.01],'',mu(i));
xM(i)=max(y(40000:end,1));
xm(i)=min(y(40000:end,1));
end
figure,plot(mu,xM,'k','linewidth',2)
hold on,plot(mu,xm,'k','linewidth',2)
plot([-2 2],[0 0],'k')
plot([0 0],[-2 2],'k')
ylabel('x'),  xlabel('\mu')
```

As theoretically predicted, system (4.1) has a stable equilibrium point for $\mu < 0$. At the bifurcation point $\mu = 0$ two branches form, representing minimum and maximum of the oscillations, that is, $\pm\sqrt{\mu}$.

4.2   Examples of oscillations and oscillators

A list of examples of oscillations and oscillators is reported:

- the day and night cycle is the most common oscillation we know;

- the movement of the Earth around the Sun is another oscillation;

- the life cycle or any process of life and death is an oscillation;

- when there are oscillators there are oscillations;

- oscillators are devices, oscillations are their effect;

- the heartbeat is controlled by an oscillator;

- the sleep and wake up cycle is an oscillation;

- oscillations allow signal propagation (Maxwell);

- the interaction between Moon and Earth leads to sea level oscillation;

- a pendulum is a mechanical oscillator.

In this chapter we discuss continuous-time oscillators. These are autonomous systems. In Chapter 2 we have discussed that, in one-dimensional maps, oscillations, including complex ones, can be generated with only one state variable. On the contrary, to obtain oscillations in a continuous-time system, at least two state variables are needed. Second-order continuous-time autonomous systems are, therefore, particularly relevant in our discussion. They are described by a set of two equations of this form:

$$\begin{aligned} \dot{x}_1 &= f_1(x_1, x_2) \\ \dot{x}_2 &= f_2(x_1, x_2) \end{aligned} \qquad (4.6)$$

where $f_1, f_2 : \mathbb{R}^2 \to \mathbb{R}$. For continuous-time systems with two state variables, the solution starting from an initial condition $\mathbf{x}(0) = [\ x_1(0) \quad x_2(0)\ ]^T = \mathbf{x}_0$ can be visualized in the plane $x_1 - x_2$ as was done for system (4.1). This plane is called the state plane or phase plane.

Therefore, for the two-dimensional case, rewritten in compact form as

$$\dot{\mathbf{x}} = f(\mathbf{x}) \qquad (4.7)$$

with $\mathbf{x} = [\ x_1 \quad x_2\ ]^T$, understanding the system behavior is facilitated by the analysis of the phase plane. For each point $\mathbf{x}$ in the plane, $f(\mathbf{x})$ is computed. As shown in Figure 4.3, the values of $f_1(\mathbf{x})$ and $f_2(\mathbf{x})$ represent the two components of $f(\mathbf{x})$. At each point $\mathbf{x}$ the vector $f(\mathbf{x})$ can be associated, it represents the direction of the flow when the system trajectory starts from $\mathbf{x}$. Drawing, for several values of $\mathbf{x}$, a vector with direction given by $f(\mathbf{x})$ and length proportional to the Euclidean norm of it, the *vector field* is built. This tool helps to visualize the direction and intensity of the flow in the phase plane. As the vector field is tangent to the trajectory in that point, the approximate trajectory of the system can be drawn by using the vector field.

The following MATLAB® commands can be used to generate the vector field of system (4.1):

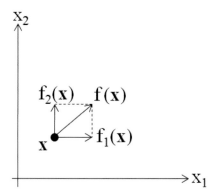

FIGURE 4.3
Construction of the vector field in the phase plane for a second-order system.

```
x10=[-2:.2:2];
x20=[-3:.2:3];

[X10,X20]=meshgrid(x10,x20);

mu=1;
dxdt1=-X20+X10.*(mu-(X10.^2+X20.^2));
dxdt2=X10+X20.*(mu-(X10.^2+X20.^2));

figure,quiver(X10,X20,dxdt1,dxdt2,2,'r')
```

In addition, we can superimpose trajectories obtained for different values of x_0 with the commands:

```
hold on
for x10=[-2:1:2]
    for x20=[-2.5:1:2.5]
        [ts,ys] = ode45('HopfOsc',[0,20],[x10,x20]);
        plot(ys(:,1),ys(:,2),'k')
    end
end
xlabel('x_1')
ylabel('x_2')
```

In doing so, we obtain the *phase portrait* of the system. The result is shown in Figure 4.4.

4.3   Genesis of electronic oscillators

This section discusses the genesis of electronic oscillators. In the last century, papers, books, and proceedings on electronic oscillators have been quite numerous. More than 500000 results are returned if the term "electronic os-

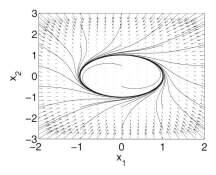

FIGURE 4.4
Vector field and a series of trajectories in the phase plane for system (4.1).

cillators" is typed in the Google search engine. As the aim of the book is to introduce the reader from simple to complex systems, providing the essential concepts in a self-contained reference, we discuss here only the main notions on electronic oscillators.

We start by considering some ideal oscillators: the lossless LC electrical oscillator, the spring mass oscillator without friction, and the mechanical pendulum, that is equivalent to the spring mass oscillator in the case of small perturbations (Figure 4.5). These systems are examples of harmonic oscillators; in fact, as we will show in the following, they oscillate around their equilibrium with a simple harmonic motion, that is, a sinusoidal oscillation with constant amplitude and frequency. In these systems the energy is conserved as there is no friction or dissipation.

For the LC oscillator, the energy initially stored in the capacitor and in the inductor is maintained, so that:

$$\frac{1}{2}Li^2 + \frac{1}{2}Cv^2 = E_0 \tag{4.8}$$

where i and v are the current in the inductor and voltage across the capacitor, respectively. Analogously, for the spring mass system

$$\frac{1}{2}kx^2 + \frac{1}{2}m\dot{x}^2 = E_0 \tag{4.9}$$

where x is the position of the mass in the mechanical system (and so $\dot{x}$ is its velocity). From the energy conservation, the motion equations derive. For the LC oscillator, one gets:

$$LC\frac{d^2i}{dt^2} + i = 0 \tag{4.10}$$

while for the spring mass system

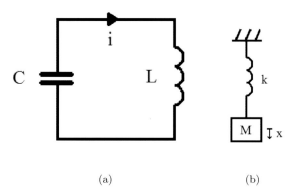

(a)   (b)

FIGURE 4.5
Examples of ideal oscillators: (a) the LC circuit; (b) the spring mass system.

$$\frac{m}{k}\frac{d^2x}{dt^2} + x = 0 \tag{4.11}$$

Both differential equations (4.10) and (4.11) have a steady-state periodic solution that is sinusoidal with frequency given by

$$\omega_0 = \frac{1}{\sqrt{LC}} \tag{4.12}$$

for the LC oscillator and

$$\omega_0 = \sqrt{\frac{k}{m}} \tag{4.13}$$

for the spring mass system.
Both systems generate undamped oscillations with amplitude depending only on the initial conditions. However, such oscillations are possible only because the ideal assumption of no damping has been made: real systems always include dissipative elements, while harmonic oscillators are ideal systems.
If dissipation is included in the electrical oscillator, the equations become

$$\frac{q}{C} + L\frac{di}{dt} + Ri = 0 \tag{4.14}$$

and so

$$\frac{i}{C} + L\frac{d^2i}{dt^2} + R\frac{di}{dt} = 0 \tag{4.15}$$

and finally

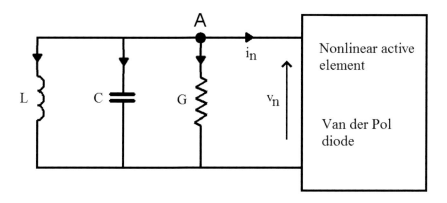

FIGURE 4.6
van der Pol circuit.

$$LC\frac{d^2i}{dt^2} + RC\frac{di}{dt} + i = 0 \qquad (4.16)$$

The differential equation (4.16) has a damped oscillatory solution. The initial energy (stored in the capacitor and in the inductor) is dissipated through the resistor $R > 0$.

The idea that is at the basis of self-sustained oscillations is to insert in a damped circuit an element compensating for the dissipation. This is obtained by using a negative driving point resistor $\bar{R} < 0$ of an active element. This principle, illustrated in Figure 4.6, underlies negative-resistance oscillators [90]. Indeed, each self-sustained oscillator includes active elements that need energy supply. The van der Pol circuit is an example of such oscillators. It includes an LC circuit, a conductance representing the losses, and a nonlinear active element, called the van der Pol diode. The nonlinear active element is supplied from the external, so to implement the circuit a supply voltage generator is also needed. Originally, the van der Pol diode was realized with a vacuum tube having nonlinear characteristics as in Figure 4.7.

The characteristic nonlinearity is that of a diode. A one-port device with this characteristic is, for instance, the tunnel diode. In fact, in the characteristic curve a region where the slope is negative (negative resistance) is evident. This negative resistance is that needed to compensate the losses due to the positive resistance $1/G$. The polarization of the one-port device is fundamental; the voltage of the power supply has to be fixed so that the diode works in the negative resistance region.

Taking into account the notation and symbols in Figures 4.6 and 4.7, we can derive the current balance at node A of the circuit as:

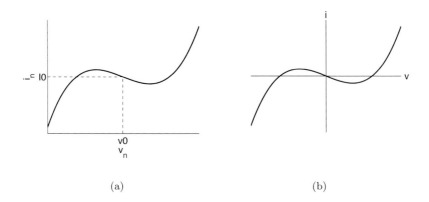

(a)   (b)

FIGURE 4.7
Characteristics of the van der Pol diode: (a) with polarization; (b) without polarization.

$$C\frac{dv_n}{dt} + Gv_n + \frac{1}{L}\int v_n dt + i_n = 0 \tag{4.17}$$

By differentiating Equation (4.17) and rearranging the terms, one gets:

$$\frac{d^2v_n}{dt^2} + \frac{1}{C}\left(G\frac{dv_n}{dt} + \frac{di_n}{dt}\right) + \frac{v_n}{LC} = 0 \tag{4.18}$$

with $i_n = F(v_n)$. Taking into account the derivative chain rule we substitute $\frac{di_n}{dt} = \frac{di_n}{dv_n}\frac{dv_n}{dt}$ into Equation (4.18):

$$\frac{d^2v_n}{dt^2} + \frac{1}{C}\left(G\frac{dv_n}{dt} + \frac{di_n}{dv_n}\frac{dv_n}{dt}\right) + \frac{v_n}{LC} = 0 \tag{4.19}$$

and so

$$\frac{d^2v_n}{dt^2} + \frac{1}{C}\frac{dv_n}{dt}\left(G + \frac{di_n}{dv_n}\right) + \frac{v_n}{LC} = 0 \tag{4.20}$$

Let us consider a polynomial approximation for the nonlinear characteristic of the van der Pol diode:

$$i_n = G(\frac{1}{3}v_n^3 - 2v_n) \tag{4.21}$$

then

$$\frac{di_n}{dv_n} = G(v_n^2 - 2) \tag{4.22}$$

and so

$$\frac{d^2 v_n}{dt^2} + \frac{1}{C} G \frac{dv_n}{dt}(v_n^2 - 1) + \frac{v_n}{LC} = 0 \tag{4.23}$$

Taking $\frac{1}{LC} = 1$ and $\frac{G}{C} = \mu$, the following differential equation representing the mathematical model of the van der Pol oscillator is obtained:

$$\ddot{x} + \mu(x^2 - 1)\dot{x} + x = 0 \tag{4.24}$$

with $x = v_n$. The van der Pol equation is a nonlinear differential equation as the van der Pol circuit is a nonlinear oscillator. Indeed, to oscillate a system must be nonlinear.

In electronics, to develop the calculations and get insights on the system working principle, the behavior of the active component is often assumed locally linear, but globally active components are always nonlinear. The van der Pol oscillator is one of the examples where the nonlinearity is fundamental to achieve an emergent behavior.

4.4 Liénard systems

The second-order nonlinear systems of the form

$$\ddot{x} + f(x)\dot{x} + g(x) = 0 \tag{4.25}$$

are called Liénard systems. The functions $f(x)$ and $g(x)$ are scalar. The van der Pol oscillator belongs to the class of Liénard systems with $f(x) = \mu(x^2 - 1)$ and $g(x) = x$.

4.4.1 The Liénard's theorem

Given a Liénard system suppose that

1. the functions $f(x)$ and $g(x)$ are continuously differentiable for all x;

2. $g(x)$ is odd ($g(-x) = -g(x)$) and positive for $x > 0$, and $f(x)$ is even ($f(-x) = f(x)$);

3. defined the function $F(x) = \int_0^x f(u)du$, there exists a single value of $x > 0$ (indicated as $x = a$) for which $F(x) = 0$, the function $F(x)$ is negative for $0 < x < a$, positive and nondecreasing for $x > a$ and $F(x) \to \infty$ as $x \to \infty$,

then, the system has a unique stable limit cycle that surrounds the origin.

The theorem, thus, provides a method to prove the existence of permanent oscillations in a system.

Example 4.2 _____

Consider the van der Pol system (4.24). It is a Liénard system with $f(x) = \mu(x^2 - 1)$ and $g(x) = x$. Note that $g(x)$ is odd and $f(x)$ is even.

Let us consider $F(x) = \int\limits_0^x f(u)du = \int\limits_0^x \mu(u^2 - 1)du = \mu\frac{x^3}{3} - \mu x$. The function $F(x)$ is equal to zero for $x = 0$ and $x = \sqrt{3}$. All the other conditions of the Liénard's theorem are satisfied if $\mu > 0$. Therefore, the van der Pol system admits a stable limit cycle for $\mu > 0$.

We note that, even if linear oscillators may have closed trajectories, they do not have limit cycles. Limit cycles are peculiar to nonlinear circuits.

4.5   Dynamics of the van der Pol oscillator

Consider again the van der Pol equation

$$\ddot{x} + \mu(x^2 - 1)\dot{x} + x = 0 \tag{4.26}$$

and the Liénard function $F(x) = \mu\frac{x^3}{3} - \mu x$. The van der Pol system may be rewritten in state space form as:

$$\begin{aligned} \dot{x} &= -F(x) + \mu y \\ \dot{y} &= -\frac{1}{\mu}x \end{aligned} \tag{4.27}$$

Exercise 4.1 _____

Prove that Equations (4.27) are the state space form of system (4.26).
Solution. To prove the result, consider the first equation of (4.27) and the function $F(x)$ and differentiate them with respect to time to obtain:

$$\begin{aligned} \ddot{x} &= -\dot{F}(x) + \mu\dot{y} \\ \dot{F}(x) &= \mu(x^2 - 1)\dot{x} \end{aligned} \tag{4.28}$$

Substituting the second equation of (4.28) into the first one and using the second equation of (4.27), we obtain:

$$\ddot{x} + \mu(x^2 - 1)\dot{x} + x = 0 \tag{4.29}$$

that is, the van der Pol equation.

The van der Pol system undergoes a Hopf bifurcation at $\mu = 0$. In fact, it has one equilibrium point $(x, y) = (0, 0)$ and the Jacobian of the system (4.27) around this equilibrium point is:

$$\mathbf{J} = \begin{bmatrix} \mu(1-x^2) & \mu \\ -1/\mu & 0 \end{bmatrix}_{(x,y)=(0,0)} = \begin{bmatrix} \mu & \mu \\ -1/\mu & 0 \end{bmatrix} \qquad (4.30)$$

Hence, the equilibrium is stable for $\mu < 0$, then becomes unstable for $\mu > 0$. At $\mu = 0$ the eigenvalues (that are complex for $|\mu| < 2$) cross the imaginary axis as it occurs in the Hopf bifurcation.

The state space equations of the van der Pol system can be further rewritten by introducing $\bar{F}(x) = \frac{x^3}{3} - x$, so that

$$\begin{aligned} \dot{x} &= \mu(y - \bar{F}(x)) \\ \dot{y} &= \frac{1}{\mu}x \end{aligned} \qquad (4.31)$$

If μ is large, $\dot{y}$ is small, and at first approximation could be assumed to be constant. In this case, the system behavior is characterized by slow-fast dynamics.

Figure 4.8(a) illustrates the slow-fast dynamics of the van der Pol when μ is large, showing the phase plane of the van der Pol system (4.31): the limit cycle is reported along with the curve $y = \bar{F}(x)$. During the slow part of the dynamics, the system trajectory follows the curve $y = \bar{F}(x)$ (from B to C and from D to A in Figure 4.8(a)), while, during the fast part, the trajectory jumps (with y almost constant) from A to B and from C to D.

The figure can be obtained by integrating system (4.31) with $\mu = 10$ and superimposing the limit cycle to the curve $y = \bar{F}(x)$. To do this, we first define the system equations in the file `VdPosc.m`:

```
function dxdt=VdPOsc(t,x)

mu=10;

dxdt=[mu*(x(2)-1/3*x(1)^3+x(1));
-x(1)/mu];

end
```

and, then, use the following commands:

```
x=[-3:0.01:3];
fx=1/3*x.^3-x;
figure,plot(x,fx,'r','linewidth',2)
hold on
[t,y] = ode45(@VdPOsc,[0,50],[1.94,0.5]);
plot(y(:,1),y(:,2),'k','linewidth',2)
xlabel('x'), ylabel('y')
```

Slow-fast dynamics is typical of many natural (earthquakes, for instance) and biological systems. A typical trend of a slow-fast variable is shown in Figure 4.8(b). This is obtained again integrating system (4.31) with $\mu = 10$ and then plotting the first state variable:

```
[t,y] = ode45(@VdPOsc,[0,50],[1.94,0.5]);
figure,plot(t,y(:,1),'k','linewidth',2)
xlabel('t'), ylabel('x')
```

On the contrary, if μ is small, one can notice that the nonlinear term

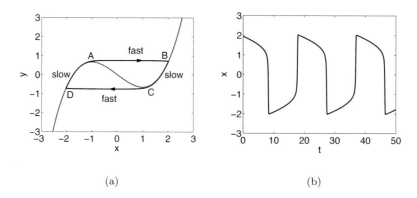

(a)   (b)

FIGURE 4.8
Slow-fast dynamics in the van der Pol system with large μ. (a) Phase plane.
(b) Typical trend of a slow-fast variable.

$\mu(x^2 - 1)$ in the van der Pol equation (4.24) is also small. Thus, under these
conditions, the system exhibits quasi-linear oscillations. This behavior is evident in Figure 4.9(a) illustrating the phase plane and, in Figure 4.9(b), reporting the waveform of x. These figures are obtained by integrating the van
der Pol equations with $\mu = 1$.

Exercise 4.2 _____

Find the phase portrait and the vector field for the van der Pol system with $\mu = 10$.
Solution. To solve the exercise, the following commands may be used:

```
x10=[-3:.2:3];
x20=[-4:.2:4];

[X10,X20]=meshgrid(x10,x20);

mu=10;
dxdt1=mu*(X20-1/3*X10.^3+X10);
dxdt2=-X10/mu;

figure,quiver(X10,X20,dxdt1,dxdt2,1,'r')
hold on
for x10=[-3:1.5:3]
    for x20=[-3:1.5:3]
        [ts,ys] = ode45('VdPOsc',[0,50],[x10,x20]);
        plot(ys(:,1),ys(:,2),'k')
    end
end
xlabel('x'), ylabel('y')
```

The result is shown in Figure 4.10. We observe that due to the large value of μ,
given a initial condition $(x(0), y(0))$, the trajectory first moves at y constant, that
is $y \simeq y(0)$, and reaches the Liénard curve, then it approaches the limit cycle.

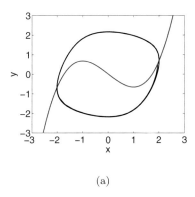 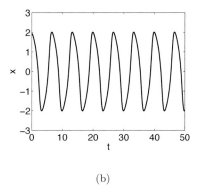

(a)   (b)

FIGURE 4.9

Dynamics of the van der Pol system for small μ. (a) Phase plane. (b) Typical trend of the van der Pol signal x for small μ.

Exercise 4.3 _____

Repeat Exercise 4.2 with $\mu = 1$.

Solution. Commands similar to those of Exercise 4.2 may be used (where the parameter μ has been changed both in the main routine and in the file VdPosc.m):

```
x10=[-3:.2:3];
x20=[-4:.2:4];

[X10,X20]=meshgrid(x10,x20);

mu=1;
dxdt1=mu*(X20-1/3*X10.^3+X10);
dxdt2=-X10/mu;

figure,quiver(X10,X20,dxdt1,dxdt2,1.5,'r')
hold on
for x10=[-3:1.5:3]
    for x20=[-3:1.5:3]
        [ts,ys] = ode45('VdPOsc',[0,50],[x10,x20]);
        plot(ys(:,1),ys(:,2),'k')
    end
end
```

to obtain the phase portrait of Figure 4.11.

Exercise 4.4 _____

By using MATLAB® find the period and frequency of oscillation of the van der Pol system with $\mu = 1$ and $\mu = 10$.

Solution. Integrating with the ode45 routine the van der Pol equations as defined in the file VdPosc.m and plotting the state variable x, the trends shown in Figure 4.12 are obtained. We measure the times, say t_1 and t_2, at which the variable x reaches two consecutive peaks, and calculate the period and frequency of oscillations. For $\mu = 1$, one gets

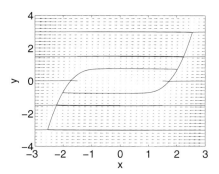

FIGURE 4.10
Phase portrait of the van der Pol system for $\mu = 10$.

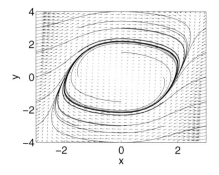

FIGURE 4.11
Phase portrait of the van der Pol system for $\mu = 1$.

$$T = t_2 - t_1 = 86.53 - 79.88 = 6.6500$$
$$f = 1/T = 0.1504$$

and for $\mu = 10$

$$T = t_2 - t_1 = 75.27 - 56.19 = 19.08$$
$$f = 1/T = 0.0524$$

Note that for $\mu = 0$ linear oscillations at a frequency equal to $f_0 = \omega_0/2\pi = 1/2\pi = 0.1592$ are obtained. Hence, when $\mu = 1$ the frequency of oscillation is very close to the linear case, confirming the effect of a weak nonlinearity.

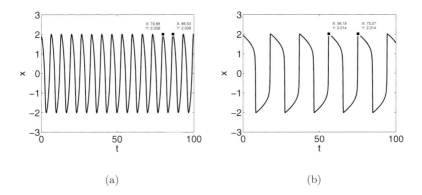

FIGURE 4.12

Trend of the state variable x of the van der Pol system for: (a) $\mu = 1$; (b) $\mu = 10$.

4.6 Lur'e systems and the design of oscillators

Lur'e systems are nonlinear systems that can be represented according to one of the two feedback schemes of Figure 4.13. Hence for systems in Lur'e form a feedback loop with a linear part $G(s)$ separated from the nonlinear one N can be established. Lur'e systems allow frequency methods to be used for their study.

Example 4.3 _____

It is easy to prove that the van der Pol equation (4.24) can be represented in Lur'e form. Consider Figure 4.14 with the linear block $G(s)$ given by $G(s) = \frac{\mu}{s^2 - \mu s + 1}$.

Taking into account that $G(s) = \frac{X(s)}{U(s)}$, if we apply the inverse Laplace transform to $X(s) = G(s)U(s)$ we obtain:

$$\ddot{x} - \mu\dot{x} + x = u$$

Now, using the fact that $u = -\mu x^2 \dot{x}$, we obtain the van der Pol equation

$$\ddot{x} + \mu(x^2 - 1)\dot{x} + x = 0$$

The frequency approach cannot be directly applied to nonlinear systems, but Lur'e systems allow us to extend some tools available for linear systems to the case of nonlinear ones. This is the case for frequency methods that work with algebraic equations instead of using nonlinear differential equations. To do this, as we will discuss below, some approximations are introduced. Such methods are physically appealing in the sense that we can expect the frequency

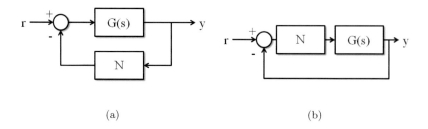

(a)   (b)

FIGURE 4.13
Systems in Lur'e form. $G(s)$ is a linear block and N is a nonlinear part.

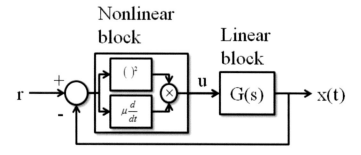

FIGURE 4.14
Lur'e form of the van der Pol oscillator.

response in accordance with the physical process knowledge. Moreover, frequency methods make use of very useful graphical tools.

4.7 Describing functions: essential elements

The fundamental idea of this method is to approximate the nonlinearity of a system with an operator that can be treated as a transfer function (Figure 4.15). Let us consider a linear time invariant system with a sinusoidal input $u(t) = A \sin(\omega t)$ and let $y(t)$ be the corresponding output. Consider the Fourier series of $y(t)$

FIGURE 4.15
Describing function to approximate a system nonlinearity.

$$y(t) = \frac{a_0}{2} + \sum_{n=1}^{\infty} [a_n \cos(n\omega t) + b_n \sin(n\omega t)] \qquad (4.32)$$

where a_i and b_i are the Fourier coefficients that, in general, depend on A and ω. If the system has some filtering properties that remove higher-order components, taking only the fundamental component $y_1(t)$ of $y(t)$ to approximate it, we have

$$y(t) \simeq y_1(t) = a_1 \cos(\omega t) + b_1 \sin(\omega t) = M \sin(\omega t + \psi) \qquad (4.33)$$

The describing function of the nonlinearity N is defined as the following complex ratio:

$$N(A, \omega) = \frac{M e^{j(\omega t + \psi)}}{A e^{j(\omega t)}} \qquad (4.34)$$

Example 4.4 _____

Determine the describing function $N(A, \omega)$ of the nonlinearity

$$y(x) = x + x^3 \qquad (4.35)$$

Solution. Given a sinusoidal input $x(t) = A \sin(\omega t)$, we have

$$\begin{aligned} y(t) &= A \sin(\omega t) + A^3 \sin^3(\omega t) = \\ &= A \sin(\omega t) + \tfrac{1}{4} A^3 [3 \sin(\omega t) - \sin(3\omega t)] \end{aligned} \qquad (4.36)$$

If we consider only the components of $y(t)$ at the fundamental frequency ω, we derive that

$$y_1(t) = A \sin(\omega t) + \frac{3}{4} A^3 \sin(\omega t) \qquad (4.37)$$

and, thus

$$N(A, \omega) = \frac{A \sin(\omega t) + \frac{3}{4} A^3 \sin(\omega t)}{A \sin(\omega t)} = 1 + \frac{3}{4} A^2 \qquad (4.38)$$

Example 4.5 _____

As a second example of describing function we examine the saturation, a nonlinearity commonly found in electronic active components. The nonlinearity, represented in Figure 4.16, is expressed by the following law:

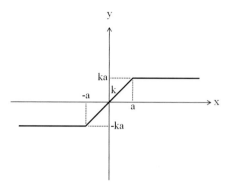

FIGURE 4.16
The saturation nonlinearity.

$$y(x) = \begin{cases} kx & \text{if } |x| \leq a \\ ka & \text{if } x > a \\ -ka & \text{if } x < -a \end{cases} \tag{4.39}$$

Consider the sinusoidal input $x(t) = A\sin(\omega t)$ with $A > a$ (in the opposite case, that is, $A < a$, the system is in the linear region), the corresponding output is given by:

$$y(t) = \begin{cases} kA\sin(\omega t) & 0 + 2h\pi < \omega t \leq \alpha + 2h\pi \\ ka & \alpha + 2h\pi < \omega t \leq \pi - \alpha + 2h\pi \\ kA\sin(\omega t) & \pi - \alpha + 2h\pi < \omega t \leq \pi + \alpha + 2h\pi \\ -ka & \pi + \alpha + 2h\pi < \omega t \leq 2\pi - \alpha + 2h\pi \end{cases} \tag{4.40}$$

with $h = 0, 1, \ldots$ and $\alpha = \sin^{-1}(a/A)$. The signal $y(t)$ is shown in Figure 4.17.
Since $y(x)$ is odd, then the even terms in the Fourier series are zero. Therefore, we only need to calculate the odd terms. As we aim to approximate $y(t)$ with the fundamental component $y_1(t) = b_1 \sin(\omega t)$ we can focus on b_1. For symmetry we can restrict its calculations only in the interval $[0, \ \pi/2]$:

$$b_1 = \frac{4}{\pi} \int_0^{\pi/2} y(t)\sin(\omega t)d(\omega t) = \frac{2kA}{\pi}\left(\alpha + \frac{a}{A}\sqrt{1 - \frac{a^2}{A^2}}\right) \tag{4.41}$$

Thus, the describing function of the saturation nonlinearity is:

$$N(A) = \frac{2k}{\pi}\left(\sin^{-1}\left(\frac{a}{A}\right) + \frac{a}{A}\sqrt{1 - \frac{a^2}{A^2}}\right) \tag{4.42}$$

The normalized function $N(A)/k$ is reported in Figure 4.18. It has been obtained with the following commands:

```
x=[1:0.05:10]; %x = A/a
y=2/pi*(asin(1./x)+1./x.*sqrt(1-(1./x).^2));
figure,plot(x,y,'k','linewidth',2)
set(gca,'FontSize',20)
xlabel('A/a'), ylabel('N(A)/k')
```

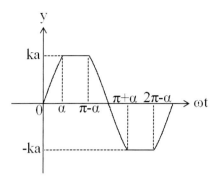

FIGURE 4.17
Output of the saturation nonlinearity when the input is $x(t) = A\sin(\omega t)$ with $A > a$.

When the describing function only depends on A, as in Example 4.5, it is indicated as $N(A)$. It can derive from single-valued nonlinearities, and in this case it is real, and from double-valued nonlinearity, e.g., hysteresis, and in this case it is complex. When the nonlinearity is dynamic, then the describing function depends on both A and ω, that is $N = N(A, \omega)$.

An exhaustive set of describing functions is reported in the book *Nonlinear Control Engineering* by Atherton [3].

Example 4.6 _____

Calculate the describing function for the nonlinearity appearing in the Lur'e form of the van der Pol oscillator.
Solution. Let $x(t) = A\sin(\omega t)$ the input of the nonlinear block. Thus, $\dot{x} = A\omega\cos(\omega t)$ and

$$y = x^2\dot{x} = \frac{A^3\omega}{2}[1 - \cos(2\omega t)]\cos(\omega t) \tag{4.43}$$

Using the trigonometric identity $\cos(\omega t)\cos(2\omega t) = \frac{1}{2}(\cos(\omega t) + \cos(3\omega t))$, one obtains

$$y(t) = \frac{A^3\omega}{4}[\cos(\omega t) - \cos(3\omega t)] \tag{4.44}$$

Taking into account that $\cos(\omega t) = \frac{1}{\omega}\frac{d}{dt}\sin(\omega t)$ and considering the fundamental component, then

$$y_1(t) = \frac{j\omega A^3}{4}\sin(\omega t) \tag{4.45}$$

and thus

$$N(A, \omega) = \frac{j\omega A^2}{4} \tag{4.46}$$

or in terms of Laplace transform

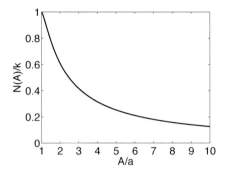

FIGURE 4.18
Output of the saturation nonlinearity when the input is $x(t) = A\sin(\omega t)$ with $A > a$.

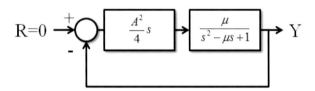

FIGURE 4.19
Block scheme of the van der Pol system in Lur'e form with describing function-based approximation of the nonlinearity.

$$N(A,s) = \frac{A^2}{4}s \qquad (4.47)$$

The nonlinearity of the van der Pol system in Lur'e form is an example of a dynamical nonlinearity and, in fact, its describing function also depends on ω.

Example 4.7 _____

Consider the block scheme of Figure 4.19 with $G(s) = \frac{\mu}{s^2-\mu s+1}$. It represents the van der Pol system in Lur'e form with the nonlinearity approximated by its describing function. We search the values of A for which the poles of the linear system in the figure are on the imaginary axis.
To this aim, consider the closed loop characteristic equation:

$$N(A,s)G(s) + 1 = 0 \qquad (4.48)$$

that for the van der Pol oscillator becomes

$$\frac{A^2}{4}s\frac{\mu}{s^2 - \mu s + 1} + 1 = 0 \qquad (4.49)$$

It follows that

$$4s^2 + s\mu(A^2 - 4) + 4 = 0 \tag{4.50}$$

Hence, the poles are on the imaginary axis if $A^2 - 4 = 0$, that is, if $A = 2$. Note that, for this value of A, $\omega = \pm j$. Therefore, both amplitude and frequency of oscillation do not depend on μ.

Clearly, the analysis based on the describing function is approximated as the higher-order components are neglected and only the fundamental component is considered in the calculations. The approximation in the final result is the cost of using a linear technique, while the major benefit is the possibility of performing the calculations in an analytical way.

More in general, in order to establish the conditions for the existence of the limit cycle with the describing function approach for system in Lur'e form, we use the closed loop characteristic equation in the frequency domain:

$$G(j\omega)N(A, \omega) + 1 = 0 \tag{4.51}$$

that yields

$$G(j\omega) = -\frac{1}{N(A, \omega)} \tag{4.52}$$

Based on Equation (4.52), to find if a system admits a limit cycle, we can draw the Nyquist plot of $G(j\omega)$ and look for intersections, if any, with the function $-\frac{1}{N(A,\omega)}$. If there exist values of A and ω for which the two curves intersect as in Figure 4.20, then a limit cycle exists.

Note that the method gives information about the existence of limit cycles, but not about their stability. In fact, to study the stability of limit cycles, analytical tools like the Loeb criterion [3] can be used. In addition, numerical simulations can help to answer the question whether a limit cycle is stable or not.

We now apply the method to the van der Pol oscillator. In this case, $G(j\omega) = \frac{\mu}{-\omega^2 - \mu j \omega + 1}$ and $-\frac{1}{N(A,\omega)} = j\frac{4}{A^2\omega}$. $G(j\omega)$ is rewritten as

$$G(j\omega) = \frac{\mu(1 - \omega^2)}{(1 - \omega^2)^2 + \mu\omega^2} + j\frac{\mu^2\omega}{(1 - \omega^2)^2 + \mu\omega^2} \tag{4.53}$$

Since $-\frac{1}{N(A,\omega)}$ is purely imaginary, an intersection may occur only if the real part of $G(j\omega)$ is zero. This occurs for $\omega = 1$. In correspondence of this frequency, the imaginary part of $G(j\omega)$ becomes

$$\frac{\mu^2\omega}{(1 - \omega^2)^2 + \mu\omega^2}\bigg|_{\omega=1} = 1 \tag{4.54}$$

Thus, we have to check that

$$\frac{4}{A^2\omega}\bigg|_{\omega=1} = 1 \tag{4.55}$$

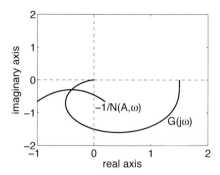

FIGURE 4.20
Illustration of the method based on the describing function to detect the existence of limit cycles.

This holds if $A = 2$.

The method is illustrated in Figure 4.21, where we have plotted the curve $-\frac{1}{N(A,\omega)}$ at fixed $A = 2$. The plotted curve is, therefore, function only of ω. This curve intersects the curve of $G(j\omega)$. As both depend on ω, one has to check that this intersection occurs at the same value of ω. This is indeed the case with $\omega = 1$, thus confirming the existence of a limit cycle.

Example 4.8 _____

Aim of this example is to show how to numerically verify the stability of a limit cycle predicted through the harmonic balance. Let us consider the feedback scheme reported in Figure 4.13(b) when $G(s) = \frac{-2s-1}{s^2+s+100}$ and the nonlinearity is a saturation as reported in Figure 4.16 with $a = k = 1$. Applying the harmonic balance approach, the existence of a limit cycle can be predicted. In Figure 4.22(a) we report the Nyquist plot of $G(j\omega)$ and, superposed the function $-\frac{1}{N(A)}$, with $N(A)$ being the describing function of the saturation as in Eqs. (4.42). The function $-\frac{1}{N(A)}$ is independent of the frequency and assumes always negative real values, therefore we look for intersections of the Nyquist plot with the real axis to predict the existence of a limit cycle. As shown in Figure 4.22(a), the intersection occurs for $-\frac{1}{N(A)} = 2$ corresponding to a frequency $\omega = 9.97$. The amplitude of the predicted limit cycle is the value of A for which $-frac1N(A) = 2$. This can be evaluated by inspecting Figure 4.22(b), where $-\frac{1}{N(A)}$ is reported as a function of A, obtaining $A \approx 2.48$.

In order to verify the stability of the predicted limit cycle, the system can be simulated defining the equations with the following commands:

```
function dxdt = system_ex(t,x)
y=x(1)+2*x(2);
h=0.5*(abs(y+1)-abs(y-1));
dxdt = [x(2);
 -100*x(1); -x(2)+h];
end
```

and integrating the dynamics through the command:

```
[T,Y]=ode45(@system_ex,[0:0.01:50],rand(2,1));
plot(T,Y(:,1)+2*Y(:,2))
```

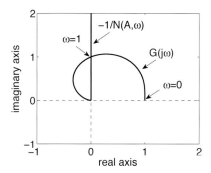

FIGURE 4.21
Illustration of the method based on the describing function to detect the existence of limit cycles for the van der Pol system.

The result is shown in Figure 4.23; it illustrates the output of the system revealing that the limit cycle is stable. The prediction is accurate, since the amplitude of the obtained limit cycle is $A \approx 2.475$, very close to the predicted one, and the frequency is $\omega = \frac{2\pi}{26.32 - 25.69} = 9.97$.

The stability of the predicted limit cycle can be also evaluated applying the *limit cycle criterion* [?], also called *Loeb criterion*. The criterion is based on the evaluation of the mutual arrangement between the Nyquist plot of $G(j\omega)$ and the function $-\frac{1}{N(A)}$ over the complex plane. The Loeb Criterion, here reported for systems withan asymptotically stable linear part $G(j\omega)$, states that at each intersection between $G(j\omega)$ and the function $-\frac{1}{N(A)}$ corresponds a limit cycle which is stable if the points near the intersection and along the curve $-\frac{1}{N(A)}$ for increasing values of A are not encircled by the curve $G(j\omega)$. Conversely, if the points near the intersection and along the curve $-\frac{1}{N(A)}$ for increasing values of A are encircled by the curve $G(j\omega)$, the predicted limit cycle is unstable.

Consider again Example 4.8 and the plot in Figure 4.22(a). We observe that, increasing A, the function $-\frac{1}{N(A)}$ assumes values not encircled by the Nyquist plot of $G(j\omega)$ thus leading to a stable limit cycle, as confirmed by the numerical simulation described therein.

4.8 Hewlett oscillator

The electrical scheme of the Hewlett oscillator is shown in Figure 4.24. An equivalent block scheme of the Hewlett oscillator in Lur'e form is now derived.

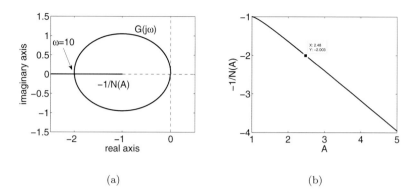

(a)   (b)

FIGURE 4.22
Harmonic balance method to predict the existence of limit cycles for the system in Example 4.8: (a) Nyquist plot of $G(j\omega)$ with superposed the function $-\frac{1}{N(A)}$, and (b) $-\frac{1}{N(A)}$ evaluated for different values of A.

Let us indicate with $Z_1(s)$ and $Z_2(s)$ the two impedances as in Figure 4.24. They are given by

$$Z_1(s) = \frac{\frac{R}{sC}}{R + \frac{1}{sC}} = \frac{R}{1 + sCR} \tag{4.56}$$

and

$$Z_2(s) = R + \frac{1}{sC} = \frac{sCR + 1}{sC} \tag{4.57}$$

We also consider $Z(s)$, that, taking into account that the operational amplifier has very high input impedance, is the series of $Z_1(s)$ and $Z_2(s)$:

$$Z(s) = \frac{R}{1 + sCR} + \frac{sCR + 1}{sC} = \frac{1 + 3sCR + s^2C^2R^2}{sC(1 + sCR)} \tag{4.58}$$

The voltage V_1 can be viewed as the result of the voltage divider formed by the two impedances $Z_1(s)$ and $Z_2(s)$, that is:

$$V_1(s) = \frac{Z_1(s)}{Z(s)} V_2(s) \tag{4.59}$$

On the other hand, it can also be obtained considering another voltage divider, this one formed by R_1 and R_2:

$$V_1(s) = \frac{R_1}{R_1 + R_2} V_2(s) \tag{4.60}$$

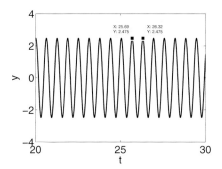

FIGURE 4.23
Output of the system in Example 4.8.

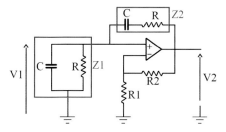

FIGURE 4.24
Electrical scheme of the Hewlett oscillator.

From Equation (4.59), we define:

$$G(s) = \frac{V_1(s)}{V_2(s)} = \frac{\frac{s}{RC}}{s^2 + \frac{3s}{RC} + \frac{1}{R^2C^2}} \tag{4.61}$$

and from Equation (4.60):

$$K = \frac{V_2(s)}{V_1(s)} = \frac{R_1 + R_2}{R_1} \tag{4.62}$$

Once calculated the transfer functions $G(s)$ and K, we can derive the block scheme equivalent to the Hewlett circuit. The scheme is shown in Figure 4.25: the signal V_2 is the input of $G(s)$ that gives as output V_1. This latter given as input of the block K produces V_2 that is fed back as input (through the voltage divider that realizes K) with opposite polarity as the feedback enters in the negative terminal of the operational amplifier.

The scheme of Figure 4.25 is linear. To take into account the nonlinearity

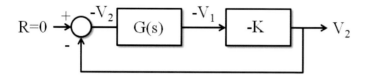

FIGURE 4.25
Feedback block scheme (linear) of the Hewlett oscillator.

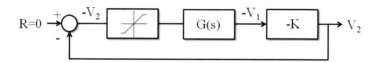

FIGURE 4.26
Feedback block scheme (nonlinear) of the Hewlett oscillator.

of the operational amplifier, a saturation block is inserted, thus obtaining the scheme of Figure 4.26.

To study the conditions for the onset of oscillations with the describing function method, the following points have to be taken into account:

- the describing function of the saturation is

$$N(A) = \frac{2k}{\pi}\left(\sin^{-1}(\frac{a}{A}) + \frac{a}{A}\sqrt{1 - \frac{a^2}{A^2}}\right) \tag{4.63}$$

- the locus of $-\frac{1}{N(A)}$ lies on the negative real axis and starts from $-\frac{1}{k}$;

- the Nyquist diagram of $KG(j\omega)$ is in the left half plane;

- the imaginary part of $KG(j\omega)$ is zero for $\omega_n = \frac{1}{RC}$.

Based on these facts, we can conclude that oscillations arise if and only if the real part of $KG(j\omega_n)$ is equal or larger than $\frac{1}{k}$, that is:

$$\text{Re}(KG(j\omega_n)) \geq \frac{1}{k}$$

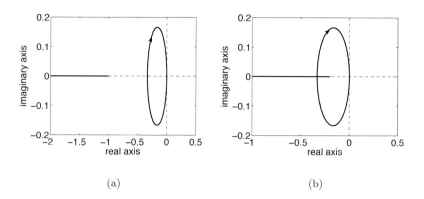

FIGURE 4.27

Illustration of the describing function approach for the Hewlett oscillator with $K = 1$, $\omega_n = 10$ and: (a) $k = 1$; (b) $k = 5$.

Example 4.9 _____

By using MATLAB®, check the existence of limit cycles for the Hewlett oscillator with different values of ω_n, k and K.

Solution. Considering the block scheme of Figure 4.26, the linear part of the Lur'e form of the Hewlett oscillator is given by $-G(s)K = -\frac{K s \omega_n}{s^2 + 3\omega_n s + \omega_n^2}$, while the non-linear part by Equation (4.63). Let's consider the following MATLAB® procedure:

```
k=1;
K=1;
wn=10;
s=tf('s');
G=-K*s*wn/(s^2+(3*wn)*s+wn^2)

A=1:0.01:10;
DFNL=2*k/pi*(asin(1./A)+sqrt(1-A.^(-2))./A);

[reG,imG] = nyquist(G);
figure,plot(squeeze(reG),squeeze(imG),'k','linewidth',2)
hold on, plot(real(-1./DFNL),imag(-1./DFNL),'k','linewidth',2)
```

where the values of the parameters have been fixed to $k = 1$, $K = 1$ and $\omega_n = 10$. As can be observed in Figure 4.27(a), there is no intersection between the Nyquist plot of $-G(s)K$ and that of $-\frac{1}{N(A)}$. Thus, for this set of parameters there is no limit cycle.

On the contrary, if the value of k in the procedure is changed to $k = 5$, then the Nyquist plot of $-G(s)K$ and that of $-\frac{1}{N(A)}$ intersect as shown in Figure 4.27(b). The intersection occurs in the point $(-0.3333, 0)$. The value of A corresponding to this point is calculated graphically. We plot in Figure 4.28 the function $-\frac{1}{N(A)}$ vs. A and graphically determine that $A \simeq 2.03$. Therefore, the approximated solution predicted by the describing function approach is $y(t) = 2.03 \sin(10t)$. We also note that, according to the Loeb criterion, the limit cycle is stable.

We remark that the describing function approach is an approximated method that, hence, gives more accurate information with weak nonlinearity.

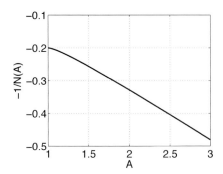

FIGURE 4.28

The function $-\frac{1}{N(A)}$ vs. A for the Hewlett oscillator.

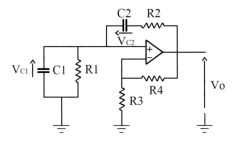

FIGURE 4.29

Electrical scheme of the Hewlett oscillator with the indication of the state variables v_{C1} and v_{C2}.

Exercise 4.5 _____

Given the Hewlett oscillator with v_{C1} and v_{C2} as in Figure 4.29, without taking into account the nonlinearity of the operational amplifier find the state space equations.
Solution. We have:

$$v_+ = v_- = \frac{R_3}{R_3 + R_4} v_0$$

and thus

$$v_0 = \frac{R_3 + R_4}{R_3} v_+ = K v_+$$

with $K = \frac{R_3 + R_4}{R_3}$. The Kirchhoff's current law yields:

$$\frac{dv_{C1}}{dt} + \frac{v_{C1}}{R_1 C_1} + \left(\frac{v_{C1}}{C_1} - \frac{v_{C2}}{C_1} - k \frac{v_{C1}}{C_1} \right) \frac{1}{R_2} = 0$$

and

$$\left(\frac{v_{C1}}{C_2} - \frac{v_{C2}}{C_2} - k\frac{v_{C1}}{C_2}\right)\frac{1}{R_2} = \frac{dv_{C2}}{dt}$$

Hence, the state space equations read

$$\begin{bmatrix} \frac{dv_{C1}}{dt} \\ \frac{dv_{C2}}{dt} \end{bmatrix} = \begin{bmatrix} -\frac{1}{R_1 C_1} - \left(\frac{1}{C_1} - \frac{k}{C_1}\right)\frac{1}{R_2} & -\frac{1}{R_2 C_1} \\ \frac{1}{R_2 C_2} - \frac{k}{R_2 C_2} & -\frac{1}{R_2 C_2} \end{bmatrix} \begin{bmatrix} v_{C1} \\ v_{C2} \end{bmatrix}$$

4.9   2D maps

In Chapter 2 we have discussed the dynamical behavior of the logistic map that is a one-dimensional map, yet very rich in dynamics. Now, we illustrate some 2D maps. This term indicates nonlinear discrete-time systems with two state variables.

In this chapter we have taken into account second-order autonomous continuous-time systems with particular emphasis on their ability to produce oscillatory behavior. The bifurcation diagrams derived for them do not show complex bifurcation features like period doubling cascades. The scenario is quite different with 2D maps that, like 1D maps, contain very rich dynamical behavior in their structural simplicity.

4.9.1   The Henon map

This map [79] was introduced by the French mathematician and astronomer Michel Henon (1931-2013) as a model of the Poincaré section of the Lorenz system that will be introduced in the next chapter. Henon also used this map to describe the orbits of celestial bodies. Henon's idea was that low-energy celestial bodies display periodic motion, while high-energy celestial bodies follow a chaotic motion.

Even if a lot of papers and experiments have been done on the Henon map and even if searching for it in Google returns about 100000 results, a short paragraph will be devoted to this map. Our aim is to stress the peculiarities of its bifurcation diagram.

The map is described by the following equations:

$$\begin{aligned} x_{k+1} &= 1 - ax_k^2 + y_k \\ y_{k+1} &= bx_k \end{aligned} \tag{4.64}$$

where a and b are constant parameters, both important to characterize the system behavior. A classical example reported in the literature of chaotic behavior is obtained by selecting $a = 1.4$ and $b = 0.3$.

Let us consider the bifurcation diagram fixing $b = 0.3$ and varying a. The MATLAB® commands to build the bifurcation diagram are the following:

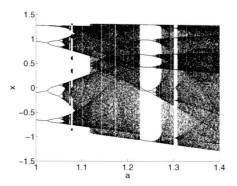

FIGURE 4.30
Bifurcation diagram of the Henon map with respect to a $(b = 0.3)$.

```
T=1000;
b=0.3;
x=zeros(T,1);
y=zeros(T,1);
x(1)=0.001;
y(1)=0.001;
figure, hold on
for a=1:0.001:1.4

    for k=1:T-1
        x(k+1)=1-a*x(k)^2+y(k);
        y(k+1)=b*x(k);
    end

    plot(a,x(600:end),'k.','MarkerSize',2)
end
xlabel('a'), ylabel('x')
```

The result is shown in Figure 4.30. We note that there are several period doubling cascades. Let us focus on the interval $a \in [1.05, 1.06]$ and calculate the Feingebaum constant by approximating it. From the diagram of Figure 4.31 we get $a_{16} = 1.05144$, $a_{32} = 1.05664$ and $a_{64} = 1.05776$ and so:

$$\delta = \frac{a_{32} - a_{16}}{a_{64} - a_{32}} = \frac{1.05664 - 1.05144}{1.05776 - 1.05664} = 4.6428 \simeq 4.6692 \qquad (4.65)$$

Exercise 4.6 _____

Find the Feigenbaum constant fixing $a = 1.1$ and using b as the bifurcation parameter for the Henon map.
Solution. The following MATLAB® commands can be used:

```
T=1000;
a=1.1;
x=zeros(T,1);
y=zeros(T,1);
x(1)=0.001;
y(1)=0.001;
```

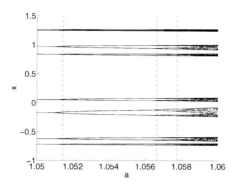

FIGURE 4.31

Magnification of the bifurcation diagram of the Henon map with respect to a ($b = 0.3$) in the interval $[1.05, 1.06]$ where a period doubling cascade occurs.

```
figure, hold on
for b=0.2:0.0005:0.4

    for k=1:T-1
        x(k+1)=1-a*x(k)^2+y(k);
        y(k+1)=b*x(k);
    end

    plot(b,x(800:end),'k.','MarkerSize',4)
end
ylim([-1.5 1.5])
xlabel('b'), ylabel('x')
```

to produce the bifurcation diagram of Figure 4.32. A period doubling cascade occurs, so we can focus on the interval $b \in [0.21, 0.27]$:

```
T=1000;
a=1.1;
x=zeros(T,1);
y=zeros(T,1);
x(1)=0.001;
y(1)=0.001;
figure, hold on
for b=0.21:0.0001:0.25

    for k=1:T-1
        x(k+1)=1-a*x(k)^2+y(k);
        y(k+1)=b*x(k);
    end

    plot(b,x(950:end),'k.','MarkerSize',4)
end
xlim([0.21 0.25])
ylim([-1.5 1.5])
xlabel('b'), ylabel('x')
```

The bifurcation diagram obtained is shown in Figure 4.33 from which we estimate the Feigenbaum constant as:

$$\delta = \frac{0.243 - 0.216}{0.249 - 0.243} = 4.5 \simeq 4.6692 \qquad (4.66)$$

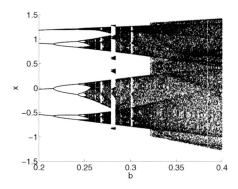

FIGURE 4.32
Bifurcation diagram of the Henon map with respect to parameter b ($a = 1.1$).

4.9.2   The Lozi map

The Lozi map [79] is another 2D discrete-time map:

$$\begin{aligned}
x_{k+1} &= 1 - a|x_k| + y_k \\
y_{k+1} &= bx_k
\end{aligned} \tag{4.67}$$

It can be obtained by the Henon map by substituting the term ax_k^2 with $a|x_k|$. The same procedure is done to the logistic map to derive the 1D Tent map.

Exercise 4.7 _____

Find the bifurcation diagrams of the Lozi map, first varying a at fixed b and, then, varying b at fixed a.
Solution. The bifurcation diagram with respect to a at fixed b ($b = 0.3$) is obtained with the following MATLAB® commands:

```
T=1000;
b=0.3;
x=zeros(T,1);
y=zeros(T,1);
x(1)=0.001;
y(1)=0.001;
figure, hold on
for a=0.6:0.001:1.6

    for k=1:T-1
        x(k+1)=1-a*abs(x(k))+y(k);
        y(k+1)=b*x(k);
    end

    plot(a,x(700:end),'k.','MarkerSize',2)
end
xlim([0.6 1.601])
set(gca,'FontSize',20)
ylabel('x'), xlabel('a')
```

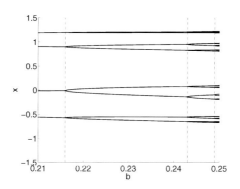

FIGURE 4.33

Magnification of the bifurcation diagram of the Henon map with respect to b ($a = 1.1$) in the interval $[0.21, \ 0.25]$ where a period doubling cascade occurs.

The result is shown in Figure 4.34. To obtain the bifurcation diagram with respect to b (with $a = 1.5$), the MATLAB® commands are:

```
T=1000;
a=1.5;
x=zeros(T,1);
y=zeros(T,1);
x(1)=0.001;
y(1)=0.001;
figure, hold on
for b=-0.6:0.002:0.6
    for k=1:T-1
        x(k+1)=1-a*abs(x(k))+y(k);
        y(k+1)=b*x(k);
    end
    plot(b,x(950:end),'k.','MarkerSize',2)
end

xlim([-0.6 0.6])
set(gca,'FontSize',20)
ylabel('x')
xlabel('b')
```

The bifurcation diagram obtained is illustrated in Figure 4.35.

4.9.3 The Ikeda map

The Ikeda map [79] is a discrete-time map with a single complex variable. Therefore, it is equivalent to a 2D map with real state variables. It is related to problems of optical turbulence and, in particular, proves the chaotic behavior of transmitted light from a ring cavity.

The map is expressed by the following iterative equation:

$$z_{k+1} = r + c_2 z_k e^{j\left(c_1 - \frac{c_3}{1+|z_k|^2}\right)} \tag{4.68}$$

with $z_k = x_k + j y_k$.

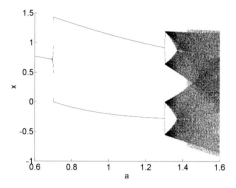

FIGURE 4.34
Bifurcation diagram of the Lozi map with respect to a ($b = 0.3$) in the interval $[0.6, \ 1.6]$.

A classical example of a set of parameters for which the Ikeda map exhibits chaotic behavior is: $c_1 = 0.4$, $c_2 = 0.9$, $c_3 = 7.2788$, and $r = 0.85$. Starting from this set of parameters, we will show, in the following, bifurcation diagrams obtained with respect to a single control parameter while the other are kept fixed. In doing so, we will also identify regions where period doubling cascades occur and estimate the Feigenbaum constant, finding a further confirmation of the universal mechanism leading to chaos.

The first bifurcation diagram, shown in Figure 4.36, is obtained by varying the parameter c_1 and keeping fixed the others to $c_2 = 0.9$, $c_3 = 7.2788$, and $r = 0.85$. The MATLAB® commands are the following:

```
T=1000;
x=zeros(T,1);
y=zeros(T,1);
x(1)=0.1;
y(1)=0.1;
z=x+i*y;

c2=0.9;
c3=7.2788;
r=0.85;

figure, hold on
for c1=-0.3:0.001:0.5

    for k=1:T-1
        z(k+1)=r+c2*z(k)*exp(i*(c1-c3/(1+abs(z(k))^2)));
    end

    plot(c1,real(z(800:end)),'k.','MarkerSize',2)
end
xlim([-0.3 0.5]), ylim([-1 2])
xlabel('c_1'), ylabel('x')
```

This diagram allows us to identify the presence of a period doubling

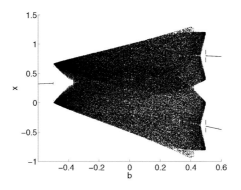

FIGURE 4.35

Bifurcation diagram of the Lozi map with respect to b $(a = 1.5)$ in the interval $[-0.6,\ 0.6]$.

cascade. Focusing on the period doublings occurring in the interval $c_1 \in [0.02,\ 0.075]$, the Feigenbaum constant can be estimated as:

$$\delta = \frac{0.0635 - 0.029}{0.071 - 0.0635} = 4.6 \simeq 4.6692 \qquad (4.69)$$

The second bifurcation diagram is obtained by varying the parameter c_2 and by keeping constant the other parameters to $c_1 = 0.4$, $c_3 = 7.2788$, and $r = 0.85$. The MATLAB® commands are:

```
T=1000;
x=zeros(T,1);
y=zeros(T,1);
x(1)=0.1;
y(1)=0.1;
z=x+i*y;

c1=0.4;
c3=7.2788;
r=0.85;

figure, hold on
for c2=0.2:0.001:1

    for k=1:T-1
        z(k+1)=r+c2*z(k)*exp(i*(c1-c3/(1+abs(z(k))^2)));
    end

    plot(c2,real(z(800:end)),'k.','MarkerSize',2)
end
xlim([0.2 1]), ylim([-1 2])
xlabel('c_2'), ylabel('x')
```

The bifurcation diagram is shown in Figure 4.37. Also in this case there is a cascade of period doublings. Focusing on the interval $c_2 \in [0.7,\ 0.735]$, the Feigenbaum constant is estimated:

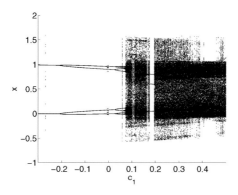

FIGURE 4.36
Bifurcation diagram of the Ikeda map with respect to c_1 (the other parameters
are fixed as $c_2 = 0.9$, $c_3 = 7.2788$, and $r = 0.85$).

$$\delta = \frac{0.7275 - 0.711}{0.731 - 0.7275} = 4.714 \simeq 4.6692 \tag{4.70}$$

The third diagram is with respect to c_3 with $c_1 = 0.4$, $c_2 = 0.9$, and
$r = 0.85$. The commands used are:

```
T=1000;
x=zeros(T,1);
y=zeros(T,1);
x(1)=0.1;
y(1)=0.1;
z=x+i*y;

c1=0.4;
c2=0.9;
r=0.85;

figure, hold on
for c3=4:0.001:8

    for k=1:T-1
        z(k+1)=r+c2*z(k)*exp(i*(c1-c3/(1+abs(z(k))^2)));
    end

    plot(c3,real(z(900:end)),'k.','MarkerSize',2)
end
xlim([4 8]),ylim([-1 2])
xlabel('c_3'), ylabel('x')
```

The estimation of the Feigenbaum constant is:

$$\delta = \frac{4.98 - 4.28}{5.13 - 4.98} = 4.7666 \simeq 4.6692 \tag{4.71}$$

The fourth bifurcation diagram is with respect to r. It is shown in Figure 4.39 and obtained with the following commands:

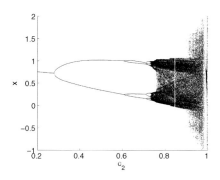

FIGURE 4.37

Bifurcation diagram of the Ikeda map with respect to c_2 (the other parameters are fixed as $c_1 = 0.4$, $c_3 = 7.2788$, and $r = 0.85$).

```
T=1000;
x=zeros(T,1);
y=zeros(T,1);
x(1)=0.1;
y(1)=0.2;
z=x+i*y;

c1=0.4;
c2=0.9;
c3=7.2788;

figure, hold on
for r=0.5:0.0005:1
    for k=1:T-1
        z(k+1)=r+c2*z(k)*exp(i*(c1-c3/(1+abs(z(k))^2)));
    end
    z(1)=z(T);

    plot(r,real(z(800:end)),'k.','MarkerSize',2)
end
xlim([0.5 1]), ylim([-1 2])
xlabel('r'), ylabel('x')
```

The estimation of the Feigenbaum constant is done in the interval $r \in [0.64, 0.84]$ and gives:

$$\delta = \frac{0.786 - 0.655}{0.814 - 0.786} = 4.618 \simeq 4.6692 \qquad (4.72)$$

We note that the same results can be obtained if the imaginary part of z is used instead of the real part. All the cases analyzed confirm the universal route to chaos governed by the Feigenbaum constant.

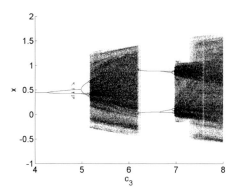

FIGURE 4.38
Bifurcation diagram of the Ikeda map with respect to c_3 (the other parameters
are fixed as $c_1 = 0.4$, $c_2 = 0.9$, and $r = 0.85$).

4.10   Summary

In this chapter some classical schemes of oscillators have been discussed. Their
study highlights that:

- oscillations are generated through Hopf bifurcation;

- a negative resistance has to be included in the circuit to obtain oscillations;

- the describing function approach can be used to design oscillators.

In addition, the chapter discusses 2D discrete-time maps showing that,
contrary to second-order continuous-time autonomous systems, their behavior
can also be chaotic.

The analysis of the systems reported in the chapter makes clear the importance of the bifurcation diagrams as a tool to characterize the dynamical
behavior of nonlinear systems. For 2D maps these diagrams confirm the universality of the Feigenbaum constant and period doubling route to chaos.

4.11   Exercises

1. Consider the system:

$$\ddot{x} + \mu(x^6 - 1)\dot{x} + x = 0$$

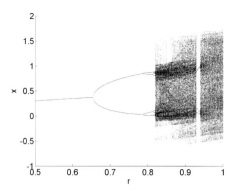

FIGURE 4.39
Bifurcation diagram of the Ikeda map with respect to r (the other parameters are fixed as $c_1 = 0.4$, $c_2 = 0.9$, and $c_3 = 7.2788$).

analyze it using the Liénard theorem and then write a MATLAB® script to draw its phase portrait.

2. Find the value of a for which the following system oscillates:

$$\dot{x} = ax - y - x^3$$
$$\dot{y} = x + ay - y^3 \tag{4.73}$$

3. The following set of differential equations represents a model of the cell cycle proposed by Tyson [88]:

$$\dot{x} = b(y - x)(\alpha + x^2) - x$$
$$\dot{y} = c - x \tag{4.74}$$

where x and y are proportional to two protein concentrations. Suppose that $b \gg 1$ and $\alpha \ll 1$, satisfying the constraint $8\alpha b < 1$. Choose the values of the parameters a and b and draw the bifurcation diagram with respect to parameter c.

4. Propose an electronic model of the system in the previous exercise.

5. Consider the following system:

$$\dot{x}_1 = x_2$$
$$\dot{x}_2 = -\alpha x_2 - \sin x_1 \tag{4.75}$$

Draw the phase portrait for various values of α, including the frictionless case $\alpha = 0$.

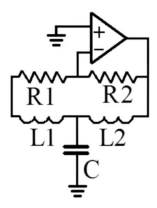

FIGURE 4.40
The Hartley oscillator.

6. Using the MATLAB® script proposed for the previous exercise, draw the phase portrait for the second-order linear system with various configurations of the eigenvalues and of the eigenvectors.

7. Consider the system:

$$\dot{x}_1 = x_2$$
$$\dot{x}_2 = -x_1 + \alpha x_2(1 - 3x_1^2 - 2x_2^2) \qquad (4.76)$$

Are there values of α for which the system oscillates?

8. Draw the phase portrait, for various values of α, of the following system:

$$\dot{x}_1 = x_2$$
$$\dot{x}_2 = x_1 - x_2(1 - x_1^2 + \alpha x_1^4) \qquad (4.77)$$

Discuss the results.

9. Consider the electronic scheme reported in Figure 4.40 representing the Hartley oscillator. Write the dynamical equations and derive the conditions for oscillations.

10. Consider the following second-order map:

$$x_{k+1} = x_k + A\sin(2\pi x_k) \quad (\mod 2\pi)$$
$$y_{k+1} = y_k + A\sin(2\pi x_k) \quad (\mod 2\pi) \qquad (4.78)$$

Derive the bifurcation diagram with respect to the parameter A.

Further reading

For additional information on the topics of the chapter, the following references may be consulted: [3], [22], [48], [78], [82], [92].

5

Strange attractors and continuous-time chaotic systems

CONTENTS

In the previous chapter periodic oscillators and oscillations have been discussed; second-order systems have been analyzed and the describing function approach to obtain approximate solutions of them has been introduced. Indeed, for autonomous nonlinear continuous-time systems aperiodic oscillations have not been observed. Moreover, this behavior has been analyzed in detail in Chapter 2 for the case of 1D maps, where the term chaos has been introduced, and in the second part of Chapter 4 for 2D maps.

In this chapter strange attractors and chaos in continuous-time systems [22, 70] are presented. As the reader can observe the route of the book is in agreement with the graph initially reported in Chapter 1 and referred to as Figure 1.3. Increasing the number of variables and including strongly nonlinear terms, more complex phenomena are observed. This chapter discusses methods and numerical experiments on continuous-time systems with chaotic attractors.

The electronic implementation and applications of chaotic elementary circuits will be the focus of Chapter 8.

5.1　Features of chaos in continuous-time systems

All the fingerprints of chaos in discrete-time systems are also retrieved in continuous-time systems. Therefore

- aperiodic oscillations of the state variables,

- high sensitivity to initial conditions,

- sensitivity to parameter changes,

- period doubling cascades, if any, ruled by the Feigenbaum constant,

- long-term unpredictability,

- signals with a wide spectrum, similar to that of white noise,

are all features of continuous-time chaotic systems as well.

Moreover, since in deterministic systems state space trajectories have no common points, aperiodicity in autonomous continuous-time systems is only possible if the number of state variables is at least three. In the case of continuous-time systems, attractors may also be *strange*. The term refers to the fact that the orbit is bounded, but not periodic or convergent; on the contrary, it has a complex fractal structure. Chaotic systems display strange attractors, that is, the trajectories are confined in a limit set, in which an infinite number of trajectories approach each other without intersecting one another. Given any point of a trajectory, at some time the trajectory will return arbitrarily close to that point, but will never pass again through the same point, thus forming a dense set of points in a specific geometric structure. Despite the high sensitivity to initial conditions that makes each trajectory unique, the unfolding of the trajectory in the phase space is a geometric structure that is qualitatively the same for each initial condition.

Example 5.1

Let us consider the following nonlinear system that was introduced by the American mathematician and meteorologist Edward Lorenz in 1963 [52], and for this reason it is named the Lorenz system. It is described by the following set of equations:

$$
\begin{aligned}
\dot{x} &= \sigma(y - x) \\
\dot{y} &= \rho x - y - xz \\
\dot{z} &= xy - bz
\end{aligned}
\tag{5.1}
$$

where σ, ρ and b are positive constants. Let us consider $\sigma = 10$, $\rho = 28$ and $b = 8/3$. Often, ρ is considered as a bifurcation parameter.

By using MATLAB®, we determine the trajectory in the three-dimensional state space and draw it. First, the equations are written in the file `lorenzeqs.m`

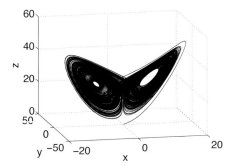

FIGURE 5.1
A trajectory of the Lorenz system.

```
function dxdt = lorenzeqs(t,x)

rho = 28;
sigma = 10;
b = 8/3;

dxdt = [sigma*(x(2) - x(1));
    x(1)*(rho - x(3)) - x(2);
    x(1)*x(2) - b*x(3)];
```

Then, they are integrated with the `ode45` routine and plotted as follows:

```
[t,x]=ode45(@lorenzeqs,[0 200],[0.1 0.1 0.2]);
figure,plot3(x(:,1), x(:,2), x(:,3))
```

The trajectory is shown in Figure 5.1.

If one imagines to cut the trajectory with a plane transverse to it and look at those points that lie on the plane, the so-called Poincaré map can be obtained. More precisely, one has to consider the intersection of an orbit in the state space of a system of order n with a subspace of order $n-1$ (that is, the Poincaré section). This subspace has to be transversal to the flow of the system as the orbit needs to flow through it. In the case of the Lorenz system the subspace is a plane. Given a point in this plane, the trajectory will leave starting from this point and then return again to this plane in another point. The Poincaré map is built by observing the first returns to the Poincaré section. It is in fact a recurrence map, that is a discrete dynamical system of lower order with respect to the original continuous-time system.

The map properties mirror those of the original system. If it is periodic, the map consists of a finite number of points, where the number of points indicates the periodicity of the motion. Otherwise, if the system is chaotic, an infinite number of points filling up a specific structure appears. The Poincaré map is thus a discrete-time system allowing us to distinguish among the different types of motion of the original system: periodic, aperiodic, chaotic.

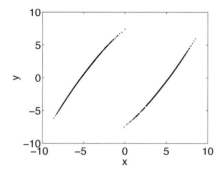

FIGURE 5.2

A Poincaré map of the Lorenz system obtained intersecting the orbit with the Poincaré section $z = 30$.

We now illustrate how to build a Poincaré map for the Lorenz attractor. In its construction, one has to pay attention to the fact that points returning on the plane with the same direction of the flow have to be selected. To this aim, the following MATLAB® function may be defined:

```
function [xp,yp,I]=poincare3D(x,y,z,zth)

z1=z(1:end-1);
z2=z(2:end);

I=find((z1>zth)&(z2<zth));

xp=x(I);
yp=y(I);

figure,plot(xp,yp,'k.')

end
```

which takes as Poincaré section the plane $z = z_{th}$. The input of the function is set of the state variables of a generic third-order system. It can be used as in this example:

```
[t,x]=ode45(@lorenzeqs,[0:0.001:1000],[0.1 0.1 0.2]);
[xp,yp]=poincare3D(x(100000:end,1),x(100000:end,2),x(100000:end,3),30);
```

that produces the plot of Figure 5.2.

There are other ways to define a discrete-time map from a continuous flow. For instance, Lorenz himself defined a map from the observation of the peaks occurring in a state variable of his system. In general, there are no systematic methods to identify the most interesting way to define a map for a continuous-time autonomous system. On the contrary, for forced continuous-time systems Poincaré maps can be defined by sampling the trajectory at the period of the forcing as discussed in the next example.

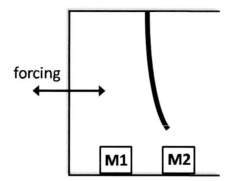

FIGURE 5.3

The Duffing system.

Example 5.2 _____

Consider a dumped Duffing equation with negative linear stiffness and external forcing:

$$\ddot{x} + \delta\dot{x} + \alpha x + \beta x^3 = \gamma\cos(\omega t) \tag{5.2}$$

The equation governs the dynamics of the system represented in Figure 5.3 that consists of a beam subjected to a mechanical forcing and whose metal tip oscillates between two magnets [59]. This example also shows that chaos can manifest in non-linear second-order forced systems. The system is a second-order non-autonomous system, and it is equivalent to a third-order autonomous system. In fact, by considering the variable time t as an additional state variable, it can be rewritten in state space form as follows:

$$\begin{aligned}
\dot{x}_1 &= x_2 \\
\dot{x}_2 &= \gamma\cos(\omega x_3) - \beta x^3 - \alpha x_1 - \delta x_2 \\
\dot{x}_3 &= 1
\end{aligned} \tag{5.3}$$

Let us consider the following values of the parameters: $\alpha = -1$, $\beta = 1$, $\gamma = 0.43$, $\delta = 0.3$, $\omega = 1.2$. The file `duffing.m` contains the system equations:

```
function dxdt = duffing(t,x)

alpha=-1;
beta=1;
gamma=0.43;
delta=0.3;
w=1.2;

dxdt=[x(2);
    gamma*cos(w*t)-alpha*x(1)-delta*x(2)-beta*x(1)^3];
```

This file is used with the following commands to generate the Poincaré map of the Duffing system:

```
w=1.2;
stepsinoneperiod=1000;
dt=2*pi/w/stepsinoneperiod;
option=odeset ('RelTol', 1e-8, 'AbsTol',1e-8);
[t,x]=ode45 (@duffing, [0:dt:10000], [0.4; 0.1],option);
```

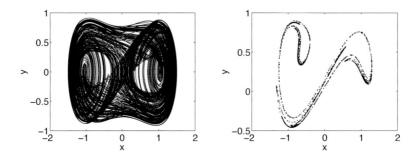

FIGURE 5.4
(a) The attractor of the Duffing system for $\alpha = -1$, $\beta = 1$, $\gamma = 0.43$, $\delta = 0.3$, $\omega = 1.2$. (b) Poincaré map of the attractor.

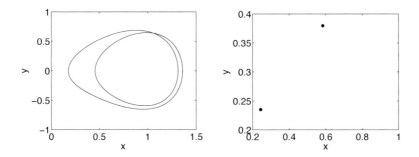

FIGURE 5.5
(a) The attractor of the Duffing system for $\alpha = -1$, $\beta = 1$, $\gamma = 0.28$, $\delta = 0.3$, $\omega = 1.2$. (b) Poincaré map of the attractor.

```
xp=x(100000:stepsinoneperiod:end,1);
yp=x(100000:stepsinoneperiod:end,2);

figure,plot(xp,yp,'k.')
xlabel('x'), ylabel('y')
```

The map is obtained by cutting the attractor with the plane $x_3 = \frac{2\pi}{\omega}$, or, recalling that the third variable represents the time, by collecting the points reached by the system trajectory sampled at regular times (given by the period of the forcing). As shown in Figure 5.4, the map has an infinite number of points; that is an indicator of deterministic chaos.

On the contrary, a period-2 limit cycle appears if the following parameters are considered: $\alpha = -1$, $\beta = 1$, $\gamma = 0.28$, $\delta = 0.3$, $\omega = 1.2$. If these parameters are inserted in the previous commands, the attractor and the Poincaré map of Figure 5.5 are obtained.

Unlike forced systems, for autonomous ones there is no trivial period to which to refer for the calculation of the Poincaré map.

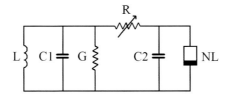

FIGURE 5.6
From the van der Pol oscillator to the Chua's circuit.

5.2   Genesis of chaotic oscillations: the Chua's circuit

In Chapter 4 permanent oscillations have been obtained by starting from a LC circuit and realizing the need of including a nonlinear element (a negative resistor) to compensate for the dissipative losses. This consideration is at the basis of the van der Pol circuit. Going forward with this reasoning we illustrate here how to build a system able to generate chaotic oscillations [46, 47]. Remember, in fact, that, being second-order and autonomous, the van der Pol equation cannot produce chaos. Something more is needed. In particular, the order of the circuit has to be increased.

Consider the scheme of Figure 5.6. With respect to the van der Pol circuit, the scheme includes a further capacitor in parallel with the nonlinear resistor and coupled with the LC circuit through a resistor R. This resistor can tune the circuit behavior: when $R \to \infty$ the circuit works as a linear LC circuit, while when $R \to 0$ the circuit reduces to a van der Pol oscillator, being the two capacitor in parallel. For intermediate values of R the circuit displays a rich dynamics: it is indeed able to generate chaotic oscillations, several limit cycles and many other nonlinear phenomena.

The circuit, which can be simplified to the scheme of Figure 5.7, was invented by Leon O. Chua and is considered as the first chaotic circuit that was intentionally designed to generate chaotic behavior [54, 31].

The equations of the circuit are:

$$\begin{array}{l} \frac{dv_1}{dt} = \frac{1}{C_1}[G(v_2 - v_1) - g(v_1)] \\ \frac{dv_2}{dt} = \frac{1}{C_2}[G(v_1 - v_2) + i_L] \\ \frac{di_L}{dt} = -\frac{1}{L}v_2 \end{array} \tag{5.4}$$

where $i = g(v_1)$ is the voltage-current characteristic of the nonlinear resistor (the explicit expression is discussed later on). With the position $x = v_1/E_1$, $y = v_2/E_1$, $z = i_L R/E_1$, $\tau = t/RC_2$, $\alpha = C_2/C_1$, $\beta = C_2 R^2/L$, and $E_1 = 1V$ they are rescaled as

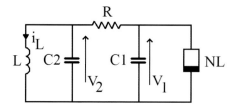

FIGURE 5.7
The Chua's circuit.

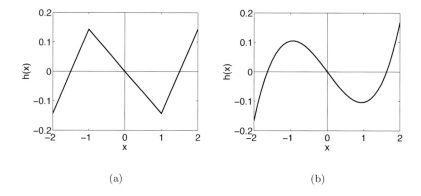

FIGURE 5.8
The nonlinearity of the dimensionless Chua's circuit: (a) piecewise linear form;
(b) polynomial form.

$$\dot{x} = \alpha[y - h(x)]$$
$$\dot{y} = x - y + z \qquad (5.5)$$
$$\dot{z} = -\beta y$$

where now $h(x)$ represents the only nonlinearity in the circuit. Equations (5.5) are usually referred to as the dimensionless Chua's circuit or, shortly, as the Chua's equations. The nonlinearity is usually expressed either as a piecewise-linear (PWL) function $h(x) = m_1 x + 0.5(m_0 - m_1)(|x + 1| - |x - 1|)$ or a continuous polynomial (cubic) expression $h(x) = c_1 x^3 + c_0 x$ as shown in Figure 5.8. The latter closely resembles the nonlinearity used in the van der Pol circuit. Typical values of the parameters of the nonlinearities are $m_0 = -1/7$, $m_1 = 2/7$, $c_1 = 1/16$, and $c_0 = -1/6$. Note that, beyond the nonlinearity often assumed with fixed parameters, only two parameters appear in the Chua's equations, namely α and β.

With reference to the piecewise-linear form of the nonlinearity, the mechanism underlying the birth of chaos in the Chua's circuit is based on a balance between the behavior in the different regions defined by the slopes of the nonlinearity. When the circuit works in the region with a negative slope, the Chua's diode provides energy to the passive network, so that oscillations are amplified. This will move the system towards one of the regions with positive slope (positive resistance), where the system is passive and oscillations are damped out, so that it returns back to the region of negative resistance and so on.

Exercise 5.1 _____

Consider the Chua's equations (5.5) with $\alpha = 9$, $\beta = 14.286$, $c_1 = 1/16$, and $c_0 = -1/6$. Numerically simulate a trajectory of the system and plot it.
Solution. The system equations are defined in the file `chuaeqs.m`:

```
function dxdt = chuaeqs(t,x)

alpha = 9;
beta = 14.286;
c1=1/16;
c0=-1/6;

dxdt = [alpha*(x(2) - c1*x(1)^3-c0*x(1));
    x(1) - x(2) + x(3);
    - beta*x(2)];
```

They are integrated with the `ode45` routine:

```
[t,x]=ode45(@chuaeqs,[0 200],[0.1 -0.2 0.1]);
```

The solution obtained is then visualized showing the evolution of each state variable (Figure 5.9):

```
figure,plot(t,x(:,1),'k')
figure,plot(t,x(:,2),'k')
figure,plot(t,x(:,3),'k')
```

The equations are then integrated for a longer time $t_f = 500$

```
[t,x]=ode45(@chuaeqs,[0 500],[0.1 -0.2 0.1]);
```

to visualize the projection of the attractor on a plane, e.g., $x-y$ or $x-z$ (Figure 5.10)

```
figure,plot(x(:,1), x(:,2),'k')
figure,plot(x(:,1), x(:,3),'k')
```

or the attractor in the state space (Figure 5.11):

```
figure,plot3(x(:,1), x(:,2), x(:,3),'k')
```

The attractor shown in Figure 5.11 is called the *double-scroll Chua's attractor*.

Example 5.3 _____

In this example we illustrate the dynamics of the Chua's equations by using the numerical approach to derive the bifurcation diagram. As done in the previous chapter for 2D maps, for the purpose of illustration we fix all the parameters but one and plot the local minima and maxima of a state variable in a diagram. The calculation of local minima and maxima is performed once the system has reached the steady-state regime. We also note that the choice of the state variable on which

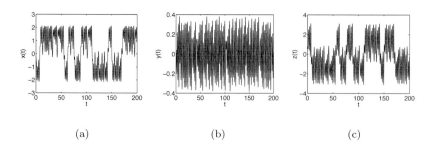

(a)   (b)   (c)

FIGURE 5.9
State variables of the dimensionless Chua's circuit with $\alpha = 9$, $\beta = 14.286$, $c_1 = 1/16$, and $c_0 = -1/6$: (a) $x(t)$; (b) $y(t)$; (c) $z(t)$.

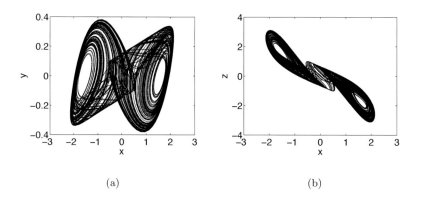

(a)   (b)

FIGURE 5.10
Projection of the attractor of the dimensionless Chua's circuit with $\alpha = 9$, $\beta = 14.286$, $c_1 = 1/16$, and $c_0 = -1/6$: (a) on the plane $x - y$; (b) on the plane $x - z$.

performing the calculations is somewhat arbitrary and similar results are obtained by selecting one of the other variables.

We first discuss the bifurcation diagram vs. α when β is kept fixed to the value $\beta = 14.286$. We rewrite the Chua's equations in the file `chuaeqs_a.m` so that the function now has a further input parameter (that is α):

```
function dxdt = chuaeqs_a(t,x,alpha)

beta = 14.286;
c1=1/16;
c0=-1/6;

dxdt = [alpha*(x(2) - c1*x(1)^3-c0*x(1));
```

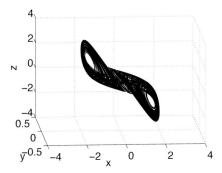

FIGURE 5.11

Chaotic attractor of the dimensionless Chua's circuit with $\alpha = 9$, $\beta = 14.286$, $c_1 = 1/16$, and $c_0 = -1/6$.

```
    x(1) - x(2) + x(3);
    - beta*x(2)];
```

We repeatedly integrate this system of equations with different values of α:

```
alpha=[6:0.005:11];
x0=[0.1 -0.2 0.1]';

figure, hold on
for i=1:length(alpha)
    [t,y]=ode45(@chuaeqs_a,[0:0.01:500],x0,'',alpha(i));
    x0=y(end,:);
    [xm,tm]=peakfind(t,y(40000:end,1));
    plot(alpha(i),xm,'k.','markersize',2)
    [xm,tm]=peakfind(t,-y(40000:end,1));
    plot(alpha(i),-xm,'k.','markersize',2)

end
```

where the function `peakfind` gives the local maxima (and times at which they occur, if needed) of a signal:

```
function [xmaxima,tmaxima] = peakfind(t,x)
xmaxima = [];
tmaxima = [];

for i=2:length(x)-1
    if (x(i-1)<=x(i))&&(x(i)>x(i+1))
        %display(i);
        tmaxima = [tmaxima; t(i)];
        xmaxima = [xmaxima; x(i)];
    end
end

end
```

Note that in the routine for the bifurcation diagram the initial condition for the next value of α is the final point of the previous numerical simulation. In fact, the Chua's circuit is a system where multiple stable attractors coexist, so if one wants to follow the bifurcation of one attractor (here one equilibrium point bifurcating to a limit cycle and so on), continuity in the numerical simulations has to be guaranteed.

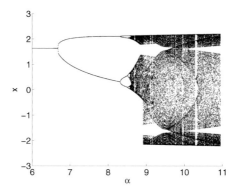

FIGURE 5.12

Bifurcation diagram vs. α of the dimensionless Chua's circuit with $\beta = 14.286$, $c_1 = 1/16$, and $c_0 = -1/6$.

The bifurcation diagram of the dimensionless Chua's circuit with respect to $\alpha = 9$ is shown in Figure 5.12. We observe in the diagram the period doubling route to chaos: for increasing values of α a stable equilibrium point bifurcates into a limit cycle and then period doubling bifurcations occur leading the system into the single-scroll chaotic attractor. This attractor in turn bifurcates into the double-scroll chaotic attractor as is clearly visible in the diagram where the values of minima and maxima spread into a larger interval of values. We also note that the system is symmetrical to the change of variables $(x, y, z) \rightarrow (-x, -y, -z)$ and, thus, up to the birth of the double-scroll chaotic attractor a symmetric attractor coexists with the one reported in the diagram.

Also in the case of the Chua's circuit, the period doubling cascade is ruled by the Feigenbaum constant. In fact, if we estimate the transition points where the period-2, period-4, and period-8 are doubled, we get:

$$\delta = \frac{8.51 - 8.3}{8.555 - 8.51} = 4.6667 \simeq 4.6692 \tag{5.6}$$

that is a good approximation of the Feigenbaum constant.

The bifurcation diagram with respect to β may be obtained following the same steps illustrated above. It is shown in Figure 5.13. We note that for low values of β the behavior becomes periodic with large amplitude oscillations. This attractor corresponds to a stable limit cycle, external to the chaotic attractor and often coexisting with it. From the period doubling cascade appearing in it, we can notice that again the estimation of the Feigenbaum constant from period doublings of period-2, period-4, and period-8 solutions

$$\delta = \frac{16.04 - 15.48}{15.48 - 15.36} = 4.6667 \simeq 4.6692 \tag{5.7}$$

provides a good approximation of the constant.

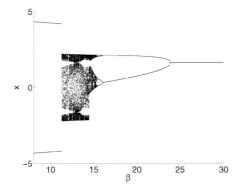

FIGURE 5.13
Bifurcation diagram vs. β of the dimensionless Chua's circuit with $\alpha = 9$, $c_1 = 1/16$, and $c_0 = -1/6$.

5.3 Canonical chaotic attractors and their bifurcation diagrams

In this section we illustrate attractors and bifurcation diagrams of some of the most studied chaotic nonlinear systems. The reader is invited to reproduce the reported diagrams and to propose further numerical experiments.

5.3.1 The Rössler system

The equations of the Rössler system [73] are

$$
\begin{aligned}
\dot{x} &= -y - z \\
\dot{y} &= x + ay \\
\dot{z} &= b + z(x - c)
\end{aligned}
\tag{5.8}
$$

It is a third-order system with the nonlinearity given by the product of two state variables. The system is chaotic for several values of its parameters. Here we fix them as $a = 0.2$, $b = 0.2$, and $c = 7$. To simulate Equations (5.8) in MATLAB® they are written in a function `rosslereqs` as follows:

```
function dxdt = rosslereqs(t,x)

a=0.2;
b=0.2;
c=7;

dxdt = [-x(2) - x(3);
    x(1)+a*x(2);
    b+x(3)*(x(1)-c)];
```

and then integrated with the command:

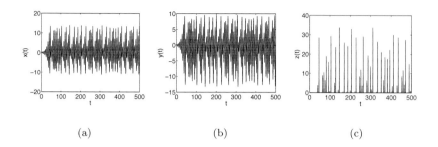

(a)   (b)   (c)

FIGURE 5.14
State variables of the Rössler system with $a = 0.2$, $b = 0.2$, and $c = 7$: (a) $x(t)$; (b) $y(t)$; (c) $z(t)$.

```
[t,x]=ode45(@rosslereqs,[0:0.01:500],[0.1 -0.25 0.15]);
```

This produces the matrix **x** containing all the calculated samples of the state variables which can be plotted as usual to inspect the trends of each state variable (Figure 5.14):

```
figure,plot(t,x(:,1),'k')
figure,plot(t,x(:,2),'k')
figure,plot(t,x(:,3),'k')
```

Figure 5.15 shows the projection of the attractor on the planes $x - y$ and $y - z$, obtained with the commands:

```
figure,plot(x(:,1), x(:,2),'k')
figure,plot(x(:,2), x(:,3),'k')
```

while the attractor in the state space is illustrated in Figure 5.16. It is obtained with the command:

```
figure,plot3(x(:,1), x(:,2), x(:,3),'k')
```

We now discuss the bifurcation diagrams of the Rössler system with respect to each single parameter appearing in the equations, that is a, b, and c. We illustrate the procedure for changes in the parameter a. The other diagrams are obtained in a similar way. First, we rewrite the MATLAB® file containing the equations of the system so that a is an input parameter for the function, that is:

```
function dxdt = rosslereqs_a(t,x)

b=0.2;
c=7;

dxdt = [-x(2) - x(3);
    x(1)+a*x(2);
    b+x(3)*(x(1)-c)];
```

This function is used in the following procedure to compute the bifurcation diagram:

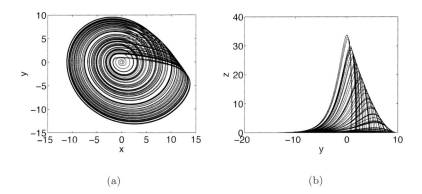

(a)   (b)

FIGURE 5.15
Projection of the attractor of the Rössler system with $a = 0.2$, $b = 0.2$, and $c = 7$: (a) on the plane $x - y$; (b) on the plane $y - z$.

```
a=[-0.1:0.0005:0.3];
x0=[0.1 -0.2 0.1]';

figure
hold on
for i=1:length(a)
    [t,y]=ode45(@rosslereqs_a,[0:0.01:500],x0,'',a(i));
    x0=y(end,:);
    [xm,tm]=peakfind(t,y(40000:end,1));
    plot(a(i),xm,'k.','markersize',2)
    [xm,tm]=peakfind(t,-y(40000:end,1));
    plot(a(i),-xm,'k.','markersize',2)

end
```

The bifurcation diagram with respect to a is shown in Figure 5.17. Those with respect to b and c are reported in Figure 5.18 and Figure 5.19, respectively.

From the bifurcation diagrams with the technique of estimating the period doublings of period-2, period-4, and period-8 solutions we can calculate approximations of the Feigenbaum constant. By doing so, we obtain:

$$\delta = \frac{0.122 - 0.0915}{0.1285 - 0.122} = 4.6923 \simeq 4.6692 \tag{5.9}$$

$$\delta = \frac{2.3 - 1.46}{1.46 - 1.28} = 4.6667 \simeq 4.6692 \tag{5.10}$$

$$\delta = \frac{4.115 - 3.81}{4.175 - 4.115} = 4.6923 \sim 4.6692 \tag{5.11}$$

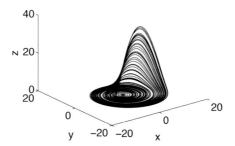

FIGURE 5.16
Chaotic attractor of the Rössler system with $a = 0.2$, $b = 0.2$, and $c = 7$.

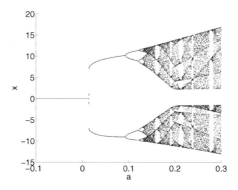

FIGURE 5.17
Bifurcation diagram vs. a of the Rössler system with $b = 0.2$ and $c = 7$.

5.3.2 The Lorenz system

The Lorenz system has already been introduced in Example 5.1; here we further discuss some remarkable properties of the system. The equations of the Lorenz system are here reported:

$$\begin{aligned}
\dot{x} &= \sigma(y - x) \\
\dot{y} &= \rho x - y - xz \\
\dot{z} &= xy - bz
\end{aligned} \tag{5.12}$$

where, as mentioned, σ, ρ, and b are the parameters. The system was derived by Edward Lorenz as an extremely simplified model of the convective motion in the atmosphere and, for this reason, such parameters have positive values. It is interesting to note that the model may be used to describe the motion of a chaotic waterwheel.

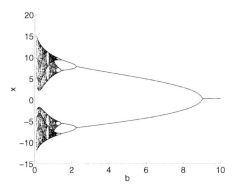

FIGURE 5.18
Bifurcation diagram vs. b of the Rössler system with $a = 0.2$ and $c = 7$.

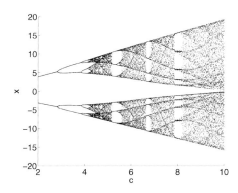

FIGURE 5.19
Bifurcation diagram vs. c of the Rössler system with $a = 0.2$ and $b = 0.2$.

We have already shown in Example 5.1 that for $\sigma = 10$, $\rho = 28$ and $b = 8/3$ chaos is obtained. By varying these parameters a variety of dynamical behaviors can be obtained. We now focus on changes of the parameter b. The bifurcation diagram with respect to this parameter can be obtained by redefining the equations in the file `lorenzeqs_b.m` so that now b can be used as an external parameter. In particular, this is implemented by including a further state variable without dynamics, that is, a constant. This alternative method is used so that some parameters for the accuracy of the integration are also passed to the `ode45` function, a step that is accomplished with the command `odeset`. The commands defining the equations to integrate are:

```
function dxdt = lorenzeqs_b(t,x)

rho = 28;
sigma = 10;
```

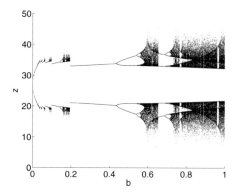

FIGURE 5.20
Bifurcation diagram vs. b of the Lorenz system with $\rho = 28$ and $\sigma = 10$.

```
dxdt = [sigma*(x(2) - x(1));
    x(1)*(rho - x(3)) - x(2);
    x(1)*x(2) - x(4)*x(3)
    0];
```

while the bifurcation diagram is calculated with the commands:

```
b=[0.002:0.001:1];
x0=[-0.0249 0.1563 0.9441]';

option=odeset ('RelTol', 1e-8, 'AbsTol',1e-8);

figure
hold on
for i=1:length(b)
    [t,y]=ode45(@lorenzeqs_b,[0:0.01:500],[x0; b(i)],option);
    [xm,tm]=peakfind(t,y(40000:end,3));
    plot(b(i),xm,'k.','markersize',2)
    [xm,tm]=peakfind(t,-y(40000:end,3));
    plot(b(i),-xm,'k.','markersize',2)

end

set(gca,'FontSize',20), ylabel('z'), xlabel('b')
```

The result is shown in Figure 5.20. It illustrates a series of period doubling cascades leading to several windows of chaos. The Feigenbaum constant rules each of these bifurcations cascades. Focusing, for instance, in the interval $b \in [0.4, \ 0.58]$, we can evaluate the following values of the critical parameter: onset of period-4 limit cycle $a_4 = 0.4302$, period-8 limit cycle $a_8 = 0.5359$, period-16 limit cycle $a_{16} = 0.5597$. From these values we derive that

$$\delta = \frac{0.5359 - 0.4302}{0.5597 - 0.5359} = 4.4412 \simeq 4.6692 \qquad (5.13)$$

Another remarkable phenomenon that can be observed in the Lorenz system is that of transient chaos [50]. This is observed for $\rho = 21$, $\sigma = 10$, and

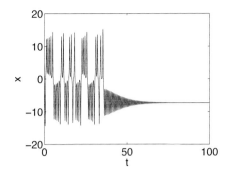

FIGURE 5.21
Transient chaos in the Lorenz system observed for $\rho = 21$, $\sigma = 10$, and $b = 8/3$. The initial condition is $(x(0), y(0), z(0)) = (9.5, -10, 9.8)$.

$b = 8/3$. For this set of parameters, two (symmetric) stable equilibrium points exist. Depending on initial conditions the trajectory ends in one of the two equilibrium points. However, for some initial conditions, before reaching one of the two equilibrium points, the system exhibits an irregular solution resembling the chaotic motion found for other values of the parameters. Figure 5.21 shows a typical trajectory characterized by transient chaos. This solution can be obtained by using the MATLAB® commands:

```
option=odeset ('RelTol', 1e-8, 'AbsTol',1e-8);
[t,x]=ode45(@lorenzeqs,[0:0.01:100],[9.5 -10 9.8],option);
figure,plot(t,x(:,1)), xlabel('t'), ylabel('x')
```

In the numerical experiments with his model Lorenz discovered a further fascinating property of the system. Given a chaotic trajectory (obtained with the usual values of the parameters $\rho = 28$, $\sigma = 10$, and $b = 8/3$), he calculated the values of the local maxima of the variable z, that is the peaks of this variable. He found a surprising regularity: let us indicate with z_k the k-th peak; if z_{k+1} is plotted as a function of z_k as in Figure 5.22, then one obtains a quite regular curve, meaning that, despite the chaoticity of the signal, we can express z_{k+1} as a function of only z_k, that is $z_{k+1} = f(z_k)$. In other words, the value of the next peak can be predicted by the knowledge of the previous peak. The function $z_{k+1} = f(z_k)$ is called the *peak-to-peak dynamics*. This surprising regularity is also found in the time intervals occurring between two successive peaks.

To compute the peak-to-peak dynamics, one has to generate a trajectory of the system, calculate the peaks and, finally, plot them. This can be done with the commands:

```
option=odeset ('RelTol', 1e-8, 'AbsTol',1e-8);
[t,y]=ode45(@lorenzeqs,[0:0.01:500],[0.5 -0.3 2],option);
[xm,tm]=peakfind(t,y(25000:end,3));
figure,plot(xm(1:end-1),xm(2:end),'k.')
```

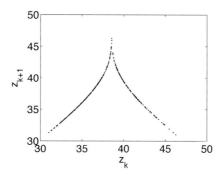

FIGURE 5.22
Peak-to-peak dynamics in the Lorenz system.

```
xlabel('z_k'),ylabel('z_{k+1}')
```

5.3.3   Thomas' cyclically symmetric attractor

The model of Thomas' cyclically symmetric attractor is described by the following equations:

$$\begin{aligned}
\dot{x} &= \sin(y) - bx \\
\dot{y} &= \sin(z) - by \\
\dot{z} &= \sin(x) - bz
\end{aligned} \tag{5.14}$$

where the parameter b indicates the amount of dissipation in the system. As one could expect, for large dissipation the system shows a stable equilibrium point, while as the parameter b is decreased a Hopf bifurcation followed by a period doubling cascade is observed. For $b = 0.18$ the system is chaotic. We simulate the equations for this value of the parameter and then derive the bifurcation diagram. To this aim, Equations (5.14) are written in a file thomaseqs.m as follows:

```
function dxdt = thomaseqs(t,x)

b=0.18;

dxdt = [sin(x(2)) - b*x(1)
    sin(x(3)) - b*x(2)
    sin(x(1)) - b*x(3)];
```

To integrate them, we use the command:

```
[t,x]=ode45(@thomaseqs,[0:0.01:500],[0.1 -0.25 0.15]);
```

and, as done for the other chaotic systems, we plot the trends of the state variables in Figure 5.23, the projection of the attractor in some plane of interest in Figure 5.24, and the attractor in Figure 5.25. The name of the attractor

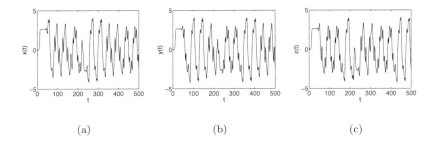

(a)   (b)   (c)

FIGURE 5.23
Evolution of the state variables of Thomas' system (5.14) with $b = 0.18$: (a)
$x(t)$; (b) $y(t)$; (c) $z(t)$.

derives from its inventor, René Thomas, and the cyclic symmetry with respect
to the variables x, y, and z of its shape.

The bifurcation diagram obtained when b is changed is illustrated in Figure 5.26, confirming that as the dissipation is decreased a rich dynamics
emerges. From the period doubling cascade of Thomas' system, the following estimation of the Feigenbaum constant is derived:

$$\delta = \frac{0.2219 - 0.2132}{0.2132 - 0.214} = 4.8333 \simeq 4.6692 \tag{5.15}$$

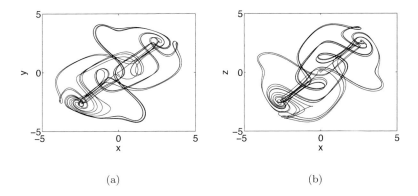

(a)   (b)

FIGURE 5.24
Projection of Thomas' cyclically symmetric attractor obtained for $b = 0.18$:
(a) $x(t)$; (b) $y(t)$; (c) $z(t)$.: (a) on the plane $x - y$; (b) on the plane $x - z$.

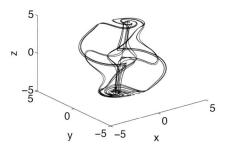

FIGURE 5.25
Thomas' cyclically symmetric attractor obtained for $b = 0.18$.

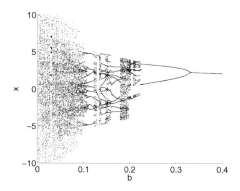

FIGURE 5.26
Bifurcation diagram vs. b of Thomas' system (5.14).

5.4　Further essential aspects of chaotic systems

The aim of this section is to give some insight into further essential aspects of chaotic systems and several tools to numerically evaluate their properties. In particular we discuss the mechanism underlying the chaos phenomenon, i.e., the interplay of two opposite effects: stretching and folding.

Let us consider a fluid in a closed tank and a heat source located below it, as schematically reported in Figure 5.27. The temperature of the particles at the bottom is higher than that of the particles closer to the top, and the same holds for their density, lower at the bottom. This triggers a mechanism so that particles at the bottom move upwards, while the particles close to the top move downwards. This is an example of the phenomenon of stretching

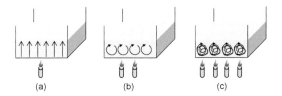

FIGURE 5.27
Heat induced hydrodynamic instability: (a) low heat source induces regular motion, (b) medium heat source induces rotational motion (Benard instability), and (c) high heat source induces irregular (chaotic) motion.

and folding in hydrodynamics as in the fluid there are opposite forces that make the particles move far away from or closer to each other. Furthermore, when the source provides a low heat (Figure 5.27(a)), the heat transfer occurs via convection, therefore mass transportation has a fundamental role. At a moderate heat source (Figure 5.27(b)) cooler particles move downwards and hotter particles move upwards inducing a regular roll motion which can be retrieved in specific areas of the fluid and which constitutes the Benard cell instability. When the heat is greater (Figure 5.27(c)), the particles have a turbulent motion characteristic of a chaotic regime. Therefore, stretching and folding lead to a mixing which tends to increase the distance of neighbor particles and back and forth to reduce the distance of initially far particles in a continuous process.

The stretching and folding phenomenon can also be schematically represented by means of the diagram reported in Figure 5.28. Let us consider a mono-dimensional string on which two points, with a relative distance δ, are marked by a circle and a square, respectively. Stretching the string and then folding it, assuming that this does not introduce a thickness in the string, implies a variation of δ. As can be noticed from Figure 5.28 after a series of stretching and folding operations on the string the distance is not only varied, but the relative positions of the two points are inverted. Further iterating this mechanism leads to an increasing unpredictability of the value of δ. The sequence of distances at each iteration is strongly influenced by the initial positions of the points, exactly as chaotic trajectories are extremely sensitive to initial conditions.

5.4.1 Computation of the Lyapunov spectrum

Lyapunov exponents have been introduced to quantify the sensitivity to perturbations on initial conditions for discrete maps in Chapter 2. The generalization for continuous-time n-th order systems is now given. Let us consider the system:

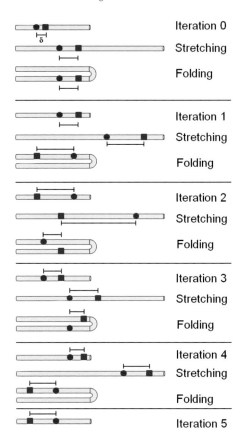

FIGURE 5.28
Stretching and folding of a mono-dimensional string.

$$\dot{\mathbf{x}} = \mathbf{f}(\mathbf{x}, t)$$
$$\mathbf{x}(t_0) = \mathbf{x}_0 \tag{5.16}$$

and a solution $\phi_t(\mathbf{x}_0, t_0)$ that represents the motion of the system starting from $\mathbf{x}_0$. Differentiating with respect to time we obtain:

$$\dot{\phi}_t(\mathbf{x}_0, t_0) = \mathbf{f}(\phi_t(\mathbf{x}_0, t_0), t) \tag{5.17}$$

with $\phi_{t_0}(\mathbf{x}_0, t_0) = \mathbf{x}_0$. If we now differentiate Equation (5.17) with respect to $\mathbf{x}_0$ we have:

$$D_{\mathbf{x}_0} \dot{\phi}_t(\mathbf{x}_0, t_0) = D_{\mathbf{x}} \mathbf{f}(\phi_t(\mathbf{x}_0, t_0), t) D_{\mathbf{x}_0} \phi_t(\mathbf{x}_0, t_0) \tag{5.18}$$

where $D_{\mathbf{x}_0} \phi_{t_0}(\mathbf{x}_0, t_0) = I$ and $D_{\mathbf{x}}$ is the Jacobian operator. We now define $\Phi_t(\mathbf{x}_0, t_0) = D_{\mathbf{x}_0} \phi_t(\mathbf{x}_0, t_0)$ and then rewrite Equation (5.19) as

$$\dot{\Phi}_t(\mathbf{x}_0, t_0) = D_\mathbf{x}\mathbf{f}(\phi_t(\mathbf{x}_0, t_0), t)\Phi_t(\mathbf{x}_0, t_0) \tag{5.19}$$

which is the so-called variational equation, a matrix-valued time-varying linear differential equation. It represents the linearization of the vector field along the trajectory $\phi_t(\mathbf{x}_0, t_0)$. Thus, the solution of the variational equation gives information on the sensitiveness of the trajectory to a perturbation on the initial conditions. In order to exemplify this concept let us consider the case of a linear time-invariant system $\dot{\mathbf{x}} = \mathbf{A}\mathbf{x}$. The corresponding variational equation can be written as $\Phi_t(\mathbf{x}_0, t_0) = AD_\mathbf{x}\dot{\Phi}_t$ with $\Phi(\mathbf{x}_0, 0) = I$. The solution of this equation is $\Phi = \mathrm{e}^\mathbf{A}\mathrm{t}$ that is a $n \times n$ time-dependent matrix. The physical meaning of Φ therefore is that each term Φ_{ij} represents the influence that a perturbation of the j-th component of the initial condition $(x_j(0))$ has in time on the i-th state variable $(x_i(t))$.

In the case of nonlinear systems, the solution of the variational equation is time-varying and depends also on the trajectory of the system at each time. Moreover, we can take into consideration as a measure of the stretching and folding $m_i(t)$, i.e., the time-varying eigenvalues of Φ. The Lyapunov exponents can be defined as:

$$\lambda_i = \lim_{t \to \infty} \frac{1}{t} \ln |e^{m_i(t)t}| \tag{5.20}$$

The set of n Lyapunov exponents is usually referred to as the Lyapunov spectrum. For chaotic systems it holds that $\lambda_1 > 0$, $\lambda_2 = 0$, $\lambda_i < 0$ with $i = 3 \dots n$. Furthermore, $\sum_{i=1}^n \lambda_i < 0$ in order to ensure that the trajectory is confined in the attractor.

The largest Lyapunov exponent represents the rate of divergence of two close trajectories. Indeed, we can have more than one positive Lyapunov exponent, and in this case we will say that the system is hyperchaotic. From a numerical point of view computing the Lyapunov exponents by following the definition in Equation (5.20) is unsuitable. Therefore it is convenient to use a procedure that, after having solved iteratively the variational equation, performs a Gram–Schmidt orthonormalization to avoid numerical drawbacks [67].

The following MATLAB® routine implements the function to calculate the Lyapunov exponents. It relies on the procedure available in [36] which implements the Gram–Schmidt orthonormalization. The function is defined as follows:

```
function [Time,Lyap_exp]=lyapunov(n,syst,init_cond,step,time)

n1=n;
n2=n1*(n1+1);
n_iter = round((time)/step);
sum_cum=zeros(n1,1);
gschmidt=zeros(n1,1);
vect_norm-zeros(n1,1);
y=zeros(n2,1);
y0=zeros(n2,1);
y(1:n)=init_cond;
```

```
for i=1:n1
    y((n1+1)*i)=rand;
end;
t=0;
lp=zeros(n1,1);
Time=0:step:(time-step);
Lyap_exp=[];
for iter=1:n_iter
    [T,Y]=ode45(syst,[t t+step],y);
    t=t+step;
    y=Y(size(Y,1),:);
    for i=1:n1
        for j=1:n1
            y0(n1*i+j)=y(n1*j+i);
        end
    end
    vect_norm(1)=0;
    for j=1:n1
        vect_norm(1)=vect_norm(1)+y0(n1*j+1)^2;
    end
    vect_norm(1)=sqrt(vect_norm(1));
    for j=1:n1
        y0(n1*j+1)=y0(n1*j+1)/vect_norm(1);
    end
    for j=2:n1
        for k=1:(j-1)
            gschmidt(k)=0;
            for l=1:n1
                gschmidt(k)=gschmidt(k)+y0(n1*l+j)*y0(n1*l+k);
            end
        end
        for k=1:n1
            for l=1:(j-1)
                y0(n1*k+j)=y0(n1*k+j)-gschmidt(l)*y0(n1*k+l);
            end
        end
        vect_norm(j)=0;
        for k=1:n1
            vect_norm(j)=vect_norm(j)+y0(n1*k+j)^2;
        end
        vect_norm(j)=sqrt(vect_norm(j));
        for k=1:n1
            y0(n1*k+j)=y0(n1*k+j)/vect_norm(j);
        end
    end
    for k=1:n1
        sum_cum(k)=sum_cum(k)+log(vect_norm(k));
    end
    for k=1:n1
        lp(k)=sum_cum(k)/(t);
    end
    Lyap_exp=[Lyap_exp lp];
    for i=1:n1
        for j=1:n1
            y(n1*j+i)=y0(n1*i+j);
        end
    end
end
```

Five inputs must be provided to the function:

- n — number of state variables;

- syst — handler of the function containing the system dynamics and its linearization;

- `init_cond` — vector containing the n initial conditions for the system trajectory;

- `step` — integration time step;

- `time_end` — final value of integration time.

Two outputs are returned by the function, namely the vector containing the temporal trends of the n Lyapunov exponents and the time at which they have been calculated. The asymptotic values are the estimated Lyapunov exponents. As concerns the function `syst` it will be discussed in the following example.

Example 5.4 _____

Compute the Lyapunov spectrum for the Chua's circuit in Equations (5.5) with $\alpha = 9$, $\beta = 14.286$, $c_0 = -1/6$, and $c_1 = 1/16$.

The MATLAB® function defining the dynamics and the corresponding linearization can be written as follows:

```
function f=syst(t,X)
alpha=9;
beta=14.286;
c1=1/16;
c0=-1/6;

x=X(1);
y=X(2);
z=X(3);

h=c1*x(1)^3+c0*x(1);

Y=[X(4), X(7), X(10);
   X(5), X(8), X(11);
   X(6), X(9), X(12)];

f=zeros(9,1);
Jac=zeros(3);

f=[alpha*(y-h);
x-y+z;
-beta*y];

Jac(1,1)=-alpha*c0-2*alpha*c1*x(1)^2;
Jac(1,2)=alpha;
Jac(1,3)=0;
Jac(2,1)=1;
Jac(2,2)=-1;
Jac(2,3)=1;
Jac(3,1)=0;
Jac(3,2)=-beta;
Jac(3,3)=0;

f(4:12)=Jac*Y;
```

In order to calculate and plot the trend of the three Lyapunov exponents, it is sufficient to run the following MATLAB® commands:

```
[T,Res]=lyapBook(3,@chua3ext,[0 1 0],0.1,150);
plot(T,Res)
```

which provides as output the plot reported in Figure 5.29. The three Lyapunov exponents are calculated as: $\lambda_1 \approx 0.3047$, $\lambda_2 \approx 0.00$, and $\lambda_3 \approx -2.97$.

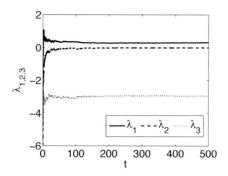

FIGURE 5.29
Lyapunov exponents calculated for the Chua's circuit.

5.4.2   The d_∞ parameter

The largest Lyapunov exponent quantifies the effect of the stretching phenomenon in a chaotic system. However, as concerns the folding, we need a further parameter. To this aim let us consider the following equation:

$$\dot{d}(t) = \lambda_1 d(t) - \gamma d(t)^2 \tag{5.21}$$

where $d(t)$ represents the distance between two trajectories of the same dynamical system starting from two different initial conditions, λ_1 is the largest Lyapunov exponent and γ is a parameter accounting for the folding phenomenon. The trend of $d(t)$ is peculiar when calculated for a chaotic system: initially, it increases with a slope equal to λ_1 until it reaches a stationary value around which it fluctuates. This stationary value is called d_∞ and can be estimated as $\frac{\lambda_1}{\gamma}$. We remark that if the system is periodic the distance between two trajectories starting from different initial conditions remains constant, while for contractive systems it tends to zero.

Example 5.5 _____

Taking into consideration the system of the previous example, draw in logarithmic scale the absolute value of the difference of the state vectors of the system calculated starting from different initial conditions.
Solution. To calculate the trend of $d(t)$ the following MATLAB® routine can be used:

```
function [d,t]=dinfchua

dinit=1e-10*rand(3,1);

x0(1)=1;
x0(2)=1.2;
x0(3)=2;

x01(1)=dinit(1)+x0(1);
x01(2)=dinit(2)+x0(2);
```

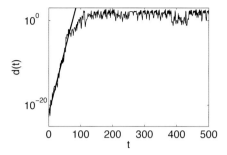

FIGURE 5.30
Calculation of the the d_∞ parameter for the Chua's circuit.

```
x01(3)=dinit(3)+x0(3);

[t,y]=ode45('chua',[0:.5:500],x0);
[t,y1]=ode45('chua',[0:.5:500],x01);
nsamples=length(y);
d=zeros(nsamples,1);

for i=1:nsamples
        delta=y(i,:)-y1(i,:);
        d(i)=log(norm(delta));
end


function yprime=chua(t,x)

alpha=9;
beta=14.286;
c1=1/16;
c0=-1/6;

h=c1*x(1)^3+c0*x(1);

yprime=[alpha*(x(2)-h);
    x(1)-x(2)+x(3);
    -beta*x(2)];
```

The semilogarithmic plot shown in Figure 5.30 represents the growing distance between the two trajectories starting from different initial conditions, which eventually reaches the d_∞ value. The growing part of the plot can be fitted through a linear polynomial whose slope is 0.2925. This value is a reliable estimation of λ_1.

The estimation of the d_∞ parameter allows us also to discern between deterministic chaos and random noise. In order to show this further feature, let us consider the following example.

Example 5.6 _____

Calculate the d_∞ for a logistic map with $a = 4$ and a random signal whose values are taken with a uniform probability distribution between 0 and 1.
Solution. Using the same code reported in the previous example, we calculate the trend of $d(t)$ for the two cases. By looking at Figure 5.31 it is clearly possible

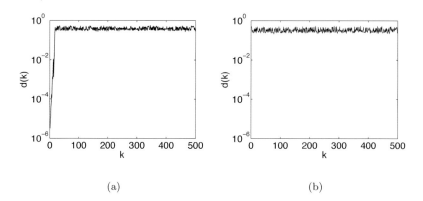

(a)   (b)

FIGURE 5.31

Trend of $d(k)$ calculated for (a) the logistic map with $a = 4$, and (b) for a random noise generator.

to observe that the two trends are sensibly different. The distance between two trajectories for the logistic map is characterized by an initial growing visible in Figure 5.31(a), leading to fluctuations around the d_∞ value, while the distance calculated between two random signals reported in Figure 5.31(b) does not show the initial growing phase. The lack of this behavior, which is peculiar to deterministic systems, allows us to discern whether or not a signal is randomly generated.

5.4.3   Peak-to-peak dynamics

Despite the fact that the unpredictability of the long-term behavior is a fingerprint of complex dynamics and of chaotic systems, a particular form of deterministic dynamics can be retrieved analyzing chaotic time-series. Let us consider the case in which a single output variable can be observed from a chaotic system and try to answer the following question: is it possible to infer the deterministic nature of the observed variable without a priori knowledge of the system? A possible answer is the analysis of the peaks of the variable and of the times at which they occur. We say that a system has a peak-to-peak dynamics (PPD) [15] if it is possible to compute the peak at time i by using the peaks that occurred at previous times as:

$$y_i = F(y_{i-1}, \ldots, y_{i-m}) \tag{5.22}$$

where y_i represents the value of the i-th peak in the time-series. The order of a PPD is equal to the number of preceding peaks needed to compute the next, i.e., $y_i = F(y_{i-1})$ is a first-order simple PPD, otherwise it is called a complex PPD. Therefore, from a continuous time dynamical system we may infer a discrete map which describes the dynamics of peaks.

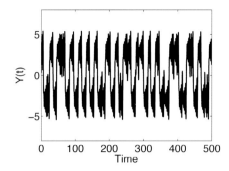

FIGURE 5.32
Trend of the observable variable generated by an unknown dynamical system.

Example 5.7 ───────────────────────────────

In Section 5.3.2 it has been shown that the Lorenz system has a peak-to-peak dynamics that displays a peculiar cusp-like shape (Figure 5.22). This peak-to-peak dynamics is first order.

5.4.4   Reconstruction of the attractor

Let us consider again the case in which we have a single observable output variable $y(t)$ of a given chaotic dynamical system. The problem referred to as *state-space reconstruction* consists of deriving information on the whole trajectory of the system, i.e., its state vector, from the output variable [42]. Assuming that the system has a given order n and that the output is sampled at constant intervals of length T, the state vector can be reconstructed as $\mathbf{x}(t) = [y(t), y(t-T), \ldots, y(t-(n-1)T)]$. The reconstructed attractor is also called an *embedding* of the original system.

The state-space reconstruction problem is indeed very hard. A good reconstruction can be obtained only having sufficient information on the dynamical system to establish the order. The choice of a suitable sampling time T is also a critical point and, in fact, a large literature exists providing different strategies to infer such information. Without entering into the many issues related to the problem of state-space reconstruction, we only discuss an example showing that the state-space reconstruction allows us to obtain an attractor from experimental data and to estimate it from a qualitative point of view.

Example 5.8 ───────────────────────────────

Consider the time-series defined by the vector Y, and reported in Figure 5.32: derive from $y(t)$ an attractor in the phase-space.
Solution. The reconstructed trajectory in phase-space fixing $T = 20$ can be obtained and plotted in MATLAB® using the following commands:

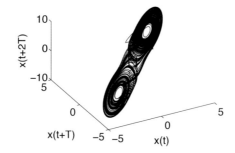

FIGURE 5.33
Three-dimensional attractor reconstruction. The state vector is reconstructed fixing $T = 20$: a double-scroll attractor appears.

```
T=20;
yr=[[Y(1:end,1)] [Y(1+T:end,1);Y(1:T,1)]
   [Y(1+2*T:end,1);Y(1:2*T,1)]];
plot3(yr(:,1),yr(:,2),yr(:,3))
```

whose output plot is reported in Figure 5.33 in which a double-scroll attractor has been reconstructed. The given time-series has been, in fact, generated integrating the Chua's circuit dynamics with the following command:

```
[t,y]==ode45(@chua,[0:0.01:100],rand(3,1));
```

and defining Y=y(:,1).

5.5   Chaotic dynamics in Lur'e systems

When does chaos arise? The answer to this question can be found in the convincing words of Mitchell Feigenbaum: "When an ordered system begins to break down into chaos, a consistent pattern of rate doubling occurs." In fact, most observations lie on the fact that "on the most irregular attractor the motion should remain unstable: trajectories diverge quickly while remaining on the attractor." Irregular attractors are generated by the combination of global compensation of phase stability with local instability of separate trajectories.

Chaos arises when oscillations become unstable; this is the general lesson that we learned from the genesis of Chua's circuit. Indeed, the mathematician René Thom tried to prove that the broken symmetries generate catastrophes, related to fundamental discontinuities, so that we can link the onset of chaos to the bifurcation of trajectories. To favor chaos, there are two opposite conditions that should be achieved: the tendency to generate oscillations and a small instability leading to turbulence and chaos.

Even if these considerations represent empirical conditions for the onset of chaotic behavior and the use of numerical tools provides reliable criteria for the study of nonlinear systems and the discovery of strange attractors, the interest of engineers is to obtain parametric conditions for the design of chaotic circuits or for predicting the bifurcation conditions for the onset of chaos.

In this section we discuss an approach to do this. In particular, the methodology is in the form of a conjecture, introduced by Roberto Genesio and Alberto Tesi in 1991 [35], and applicable to Lur'e systems. Quoting their words the conjecture says that "a Lur'e feedback system presents a chaotic behavior when a predicted limit cycle and an equilibrium point of certain characteristics interact between themselves with a suitable filtering effect along the system." In fact, chaotic motion is seen as the result of a perturbation of a periodic orbit due to some other property of the system. Clearly, this represents a basic mechanism leading to chaos and not excluding that more complex chaotic motions can be obtained by the interaction of other characteristics of the system.

More specifically, according to this conjecture a system in Lur'e form admits chaotic behavior if:

a) a stable limit cycle exists;

b) a separate unstable equilibrium point exists;

c) an interaction between the limit cycle and the equilibrium point occurs;

d) a filtering effect of the linear part of the system is present.

To apply the conjecture, first a search of a limit cycle must be performed. Assume that a steady-state solution of the closed loop system is

$$y_0(t) = A + B\sin(\omega t) \qquad (5.23)$$

with $B > 0$ and $\omega > 0$. This step can be done by using the harmonic balance method based on describing functions, in a similar way as done in Chapter 4.

The second step requires us to find an equilibrium point (EP) that is unstable. This step is accomplished analytically by linearizing the system and checking that the Jacobian matrix has at least one eigenvalue with positive real part. This equilibrium point has to be separate from the limit cycle. In fact, the essential mechanism underlying the conjecture is the interaction of two separate dynamical entities. This leads us to exclude, for instance, the equilibrium point from which the predicted limit cycle originates through a Hopf bifurcation.

The third point of the conjecture implies that the maximum amplitude of the predicted limit cycle must be greater than the equilibrium point E. Consider the output y of the system and the projection of the predicted limit cycle on the axis of this variable. If this projection does not include the equilibrium

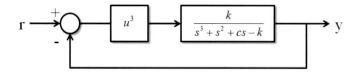

FIGURE 5.34

A system in Lur'e form with nonlinear block $N = u^3$ and linear part $G(s) = \frac{k}{s^3+s^2+cs-k}$.

point $y = E$, then no interaction is possible as the presence of the equilibrium point does not affect the trajectory of the limit cycle. On the contrary, the limit cycle will be perturbed by the unstable equilibrium point. Two forces will act in opposite directions: the perturbation by the equilibrium point moving the trajectory away from the limit cycle and the stability of the limit cycle that causes the perturbed trajectory to be attracted by the predicted limit cycle itself.

If the limit cycle and the equilibrium point do not interact with each other in this manner, the presence of the equilibrium point is not relevant and the orbit of the predicted limit cycle is not perturbed. The continuous interaction between the two dynamical objects is the factor that can lead to a chaotic strange attractor. The condition for the interaction can be expressed as:

$$A + B > E \tag{5.24}$$

and thus

$$B > |E - A| \tag{5.25}$$

Finally, the filtering effect of the linear part is a condition required by the correct application of the describing function method used to find the predicted limit cycle.

Example 5.9 ─────────────────────────────────────

Consider the Lur'e system in Figure 5.34 with nonlinearity $f(u) = u^3$ and linear part $G(s) = \frac{k}{s^3+s^2+cs-k}$. Apply the Genesio and Tesi conjecture to find if chaotic motion is possible in this system.

Solution. We follow the steps of the Genesio and Tesi conjecture.

1. Let us verify the existence of a limit cycle. To do this, we consider $u(t) = A + B\sin(\omega t)$ which leads to the so-called dual-input describing function (DIDF). In this case the output of the nonlinear block will be:

$$\tilde{y}(t) = \tilde{Y}_0 + \tilde{Y}_1 \sin(\omega t + \phi) \tag{5.26}$$

So, due to the presence of the bias term A, we define the DIDF with two components

$$N_0 = \frac{\check{Y}_0}{A}$$
$$N_1 = \frac{\check{Y}_1 e^{j\phi}}{B}$$

(5.27)

For the cubic nonlinearity we have that

$$(A + B\sin(\omega t))^3 \simeq A^3 + \frac{3}{2}AB^2 + (3A^2B + \frac{3}{4}B^3)\sin(\omega t)$$

(5.28)

where, having assumed the filtering condition, the second and third harmonics have been neglected.

Therefore, the following DIDF is obtained:

$$N_0 = A^2 + \frac{3}{2}B^2$$
$$N_1 = 3A^2 + \frac{3}{4}B^2$$

(5.29)

Note that both N_0 and N_1 are real and positive, so that $-1/N_0$ and $-1/N_1$ are real and negative.

We have now to check the condition $N(A, B)G(j\omega) = -1$ for both $N(A, B) = N_0$ and $N(A, B) = N_1$. To do this we note that $G(j\omega)$ is real for $\omega = 0$ and for $\omega = \sqrt{c}$. In particular, $G(0) = -1$ and $G(j\sqrt{c}) = -\frac{k}{k+c}$. Therefore, we have that:

$$A^2 + \frac{3}{2}B^2 = 1$$
$$(3A^2 + \frac{3}{4}B^2)k = k + c$$

(5.30)

Solving for B we obtain

$$B = 2\sqrt{\frac{2k - c}{15k}}$$

(5.31)

and

$$A = \pm\sqrt{\frac{k + 2c}{5k}}$$

(5.32)

Hence, the predicted limit cycle is characterized by $A = \pm\sqrt{\frac{k+2c}{5k}}$, $B = 2\sqrt{\frac{2k-c}{15k}}$ and $\omega = \sqrt{c}$.

2. The equilibrium point is obtained by writing the state space equations of the Lur'e system:

$$\dot{x}_1 = x_2$$
$$\dot{x}_2 = x_3$$
$$\dot{x}_3 = kx_1 - cx_2 - x_3 - kx_1^3$$

(5.33)

with $y = x_1$.

The system has three equilibrium points: $(0, 0, 0)$, $(1, 0, 0)$, and $(-1, 0, 0)$. With reference to the output variables, these points are shortly indicated as $y = 0$ and $y = \pm1$. If we consider the positive solution for A, then the condition for the interaction has to be satisfied by the separate equilibrium point $y = 1$ (in fact, $y = 0$ is not separate, and $y = -1$ is farther from the limit cycle than $y = 1$).

Linearizing the system around the equilibrium point $x = 1$, the following characteristic polynomial is obtained:

$$\lambda^3 + \lambda^2 + c\lambda + 2k = 0$$

(5.34)

For $k > 0$, the equilibrium point is unstable if $2k > c$ (this can be derived by applying for instance the Routh-Hurwitz stability criterion).

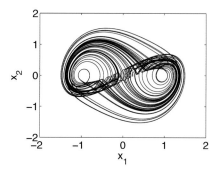

FIGURE 5.35
Projection on the plane $x_1 - x_2$ of the attractor of the Lur'e system of Figure 5.34 with $c = 1.25$ and $k = 1.8$.

3. The condition for the interaction between the equilibrium point $E = 1$ and the limit cycle is $A + B > E$, that is:

$$\sqrt{\frac{k+2c}{5k}} + 2\sqrt{\frac{2k-c}{15k}} > 1 \qquad (5.35)$$

Note that if $2k > c$, the inequality (5.35) is always satisfied.

Summarizing, the condition for the onset of chaos in the Lur'e system in Figure 5.34 is that $2k > c$.

Exercise 5.2 _____

Fix c and k such that $2k > c$ and numerically find the attractor for the system in Figure 5.34.
Solution. Let us select $c = 1.25$ and $k = 1.8$. Consider the function `lureexample`:

```
function dxdt = lureexample(t,x)

c=1.25;
k=1.8;

dxdt = [x(2)
    x(3)
    k*x(1)-c*x(2)-x(3)-k*x(1)^3];
```

and integrate it with the command:

```
[t,x]=ode45(@lureexample,[0:0.01:500],[0.1 -0.25 0.15]);
```

The attractor obtained is chaotic; its projection on the plane $x_1 - x_2$ is shown in Figure 5.35.

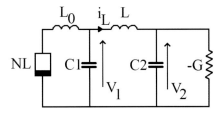

FIGURE 5.36
The Saito hyperchaotic circuit.

5.6 Hyperchaotic circuits

In this chapter, we have seen several examples of continuous-time third-order systems exhibiting chaotic dynamics for some set of parameters. Third-order models are the lowest possible order continuous-time systems where chaos can be observed, but it commonly appears also in higher-order systems. In addition, higher-order systems may also display chaotic attractors with more than one positive Lyapunov exponent. As we have seen in Section 5.4, this behavior is called hyperchaos.

As an example of hyperchaotic oscillator we discuss here a circuit due to Toshimichi Saito [74] and shown in Figure 5.36. The system is modelled by the following dimensionless equations:

$$
\begin{aligned}
\dot{x} &= -z - w \\
\dot{y} &= \gamma(2\delta y + z) \\
\dot{z} &= \rho(x - y) \\
\dot{w} &= \tfrac{1}{\varepsilon}(x - h(w))
\end{aligned}
\tag{5.36}
$$

with

$$
h(w) = \begin{cases}
w - (1 + \eta) & \text{if } w \geq \eta \\
-\eta^{-1}w & \text{if } |w| < \eta \\
w + (1 + \eta) & \text{if } w \leq -\eta
\end{cases}
\tag{5.37}
$$

This system exhibits various nonlinear phenomena including several types of chaotic motion. To have some insights of such rich behavior and on the onset of hyperchaos, the Saito oscillator is investigated by considering δ as bifurcation parameter. We first illustrate an example of hyperchaotic motion obtained as in Figure 5.37 with $\gamma = 1$, $\rho = 14$, $\varepsilon = 0.01$, $\eta = 1$, $\delta = 0.94$. This is obtained by defining the equations in the function `saitoeqs_d`:

```
function dxdt = saitoeqs_d(t,x,delta)

gamma=1;
```

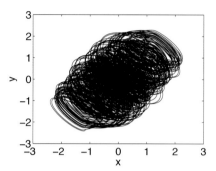

FIGURE 5.37

Projection on the plane $x - y$ of the hyperchaotic attractor of system (5.36) with $\gamma = 1$, $\rho = 14$, $\varepsilon = 0.01$, $\eta = 1$, $\delta = 0.94$.

```
rho=14;
eps=0.01;
eta=1;

w=x(4);
if w >= eta, h=w-(1+eta);
elseif w <= -eta, h=w+(1+eta);
else h=-w/eta;
end

dxdt = [-x(3)-x(4);
    gamma*(2*delta*x(2)+x(3));
    rho*(x(1)-x(2));
    1/eps*(x(1)-h)];
```

and integrating them with the command:

```
[t,x]=ode45(@saitoeqs_d,[0 500],[0.5 -0.34 1 -0.1],'',0.94);
```

Next, we study the bifurcations with respect to δ. To this purpose, the following commands may be used:

```
delta=[0.5:0.0005:1];
x0=[0.5 -0.34 1 -0.1]';

figure
hold on
for i=1:length(delta)
    i
    [t,y]=ode45(@saitoeqs,[0:0.01:500],x0,'',delta(i));
    x0=y(end,:);
    [xm,tm]=peakfind(t,y(40000:end,1));
    plot(delta(i),xm,'k.','markersize',2)
    [xm,tm]=peakfind(t,-y(40000:end,1));
    plot(delta(i),-xm,'k.','markersize',2)

end

set(gca,'FontSize',20)
ylabel('x')
xlabel('delta')
```

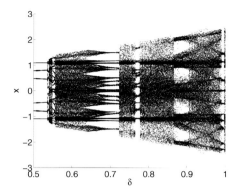

FIGURE 5.38
Bifurcation diagram with respect to δ of the Saito oscillator (5.36). The other parameters are fixed as $\gamma = 1$, $\rho = 14$, $\varepsilon = 0.01$, and $\eta = 1$.

The bifurcation diagram is shown in Figure 5.38. To clearly identify the regions of hyperchaotic motion, one has to calculate the Lyapunov exponents. Figure 5.39 illustrates the first three Lyapunov exponents (the smallest one is not reported as it is much smaller than the others). Hyperchaos appears when two Lyapunov exponents are positive. The MATLAB® commands used for the calculation of the Lyapunov exponents are the following:

```
[T,Res]=lyapunov(4,@saitoext,rand(4,1),0.01,2000);
figure,plot(T,Res)
```

which use the following functions:

```
function f=saitoext(t,X)
gamma=1;
rho=14.0;
epsi=0.01;
eta=1.0;

x=X(1);
y=X(2);
z=X(3);
w=X(4);

Y=[X(5), X(9), X(13);
   X(6), X(10), X(14);
   X(7), X(11), X(15);
   X(8), X(12), X(16)];

f=zeros(20,1);
Jac=zeros(4);

if (w>=eta)
    h=w-(1+eta);
        Jac(1,1)=0;
    Jac(1,2)=0;
    Jac(1,3)=-1;
        Jac(1,4)=-1;
    Jac(2,1)=0;
    Jac(2,2)=2*gamma*delta;
```

```
      Jac(2,2)=gamma;
      Jac(2,4)=0;
      Jac(3,1)=rho;
      Jac(3,2)=-rho;
      Jac(3,3)=0;
         Jac(3,4)=0;
      Jac(4,1)=1/epsi;
      Jac(4,2)=0;
      Jac(4,3)=0;
      Jac(4,4)=-1/epsi;
elseif (w<=-eta)
   h=w+(1+eta);
         Jac(1,1)=0;
   Jac(1,2)=0;
   Jac(1,3)=-1;
       Jac(1,4)=-1;
   Jac(2,1)=0;
   Jac(2,2)=2*gamma*delta;
   Jac(2,2)=gamma;
   Jac(2,4)=0;
   Jac(3,1)=rho;
   Jac(3,2)=-rho;
   Jac(3,3)=0;
       Jac(3,4)=0;
   Jac(4,1)=1/epsi;
   Jac(4,2)=0;
   Jac(4,3)=0;
   Jac(4,4)=-1/epsi;
elseif (abs(w)<eta)
       h=-w/eta;
       Jac(1,1)=0;
   Jac(1,2)=0;
   Jac(1,3)=-1;
       Jac(1,4)=-1;
   Jac(2,1)=0;
   Jac(2,2)=2*gamma*delta;
   Jac(2,2)=gamma;
   Jac(2,4)=0;
   Jac(3,1)=rho;
   Jac(3,2)=-rho;
   Jac(3,3)=0;
       Jac(3,4)=0;
   Jac(4,1)=1/epsi;
   Jac(4,2)=0;
   Jac(4,3)=0;
   Jac(4,4)=1/(eta*epsi);
end

f=[-z-w;
gamma*(2*delta*y+z);
rho*(x-y);
1/epsi*(x-h)];

f(5:20)=Jac*Y;
```

5.7   Summary

In this chapter further details on the chaotic behavior of continuous-time systems have been given. The Poincaré map has been introduced. The Chua's

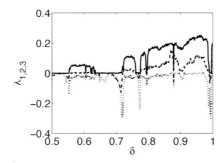

FIGURE 5.39

The first three Lyapunov exponents vs. δ of the Saito oscillator (5.36). The other parameters are fixed as $\gamma = 1$, $\rho = 14$, $\varepsilon = 0.01$, and $\eta = 1$.

circuit has been discussed, illustrating the dynamical richness of the circuit despite its simplicity in terms of number of components required for its implementation, all elements that contribute to make it the subject of many experiments.

Moreover, we remarked on the importance of bifurcation diagrams that also allow the calculation of the rate of period doubling indicating possible chaotic behavior. The numerical method to find bifurcation diagrams has been stressed in order to invite the reader to realize this type of experiment to characterize the system behavior. The aim of this chapter has also been to introduce the reader to understanding chaos and its manifestations.

Moreover, an analytical procedure based on the conjecture of Genesio and Tesi has been discussed. It can be applied to the class of Luré systems that also includes the Chua's circuit, the Lorenz system, and the Rössler system.

Hyperchaotic systems have also been briefly presented.

5.8 Exercises

1. In Exercise 6 of the previous chapter the various pole configurations have been discussed to characterize the dynamics of a second-order linear system. Repeat the exercise for the pole configurations shown in Figure 5.40 and related to third-order systems.

   The configuration $e)$ is said to be a saddle-focus configuration. Here we have coexistence of a stabilizing dynamics with a destabilizing one, so there is conflicting behavior. In the linear case the trajectory tends to infinite; in the nonlinear system the conflicting action can lead to a motion consisting

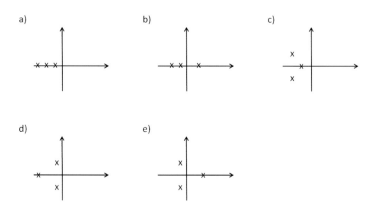

FIGURE 5.40
Pole configurations for a third-order system.

of a periodic succession of unstable trajectories towards the fixed points. Try to emphasize this concept with a suitable trajectory representation.

2. Consider the following third-order system:

$$\begin{aligned}
\dot{x}_1 &= -x_2 - x_3 \\
\dot{x}_2 &= x_1 + ax_2 \\
\dot{x}_3 &= bx_1 - cx_3 + x_1 x_3
\end{aligned} \qquad (5.38)$$

which has $(x_1, x_2, x_3) = (0, 0, 0)$ as a fixed point. Find the characteristic equation of the Jacobian and the condition for a saddle-focus. This condition is important as in the presence of a saddle-focus, a closed trajectory, called a homoclinic trajectory, starting to the equilibrium point and returning to it, can be found.

3. For the system in the previous point, verify numerically that for $a = 0.32$, $b = 0.30$, and $c = 4.50$ a chaotic attractor emerges.

4. Given the forced pendulum with equation:

$$\ddot{\theta} + b\dot{\theta} + \sin\theta = F\cos(\omega t) \qquad (5.39)$$

find b, F, and ω such that it has a chaotic behavior.

5. Consider the following non-autonomous system:

$$\begin{aligned}
\dot{x}_1 &= x_2 \\
\dot{x}_2 &= -\delta(1 + \xi\cos\omega t)x_1 - \mu x_2 - \alpha x^3
\end{aligned} \qquad (5.40)$$

with $\delta = 5$, $\xi = 14$, $\omega = 2$, $\alpha = 1$. As for the Duffing oscillator here the forcing is included in a time varying coefficient. Find the trajectory and the phase portrait for selected values of μ.

6. Consider the Lorenz system and numerically integrate its dynamics. Choosing the correct sampling period T, reconstruct the Lorenz attractor starting from the time series obtained for the x variable.

7. Consider the system:

$$\dot{x}_1 = \mu x_2 - a x_1$$
$$\dot{x}_2 = x_1 x_3 - x_2 \qquad\qquad (5.41)$$
$$\dot{x}_3 = 1 - x_1 x_2 - x_3$$

This model has been introduced by Vallis in 1988 [89] to model the temperature variation in the Earth in the equatorial area of the ocean. x_1 is the speed of the water motion, $x_2 = (T_1 - T_2)/2$ and $x_3 = (T_1 + T_2)/2$, where T_1 and T_2 are the western and the eastern ocean temperature.

(a) Study the stability of the fixed points.

(b) Fix $a = 5$ and derive the bifurcation diagram with respect to the parameter μ.

8. Discover in the Lorenz system the existence of a saddle-node and saddle-focus interactions. The curve connecting two saddle-focus points is a heteroclinic trajectory.

9. Consider the Chua's circuit dynamics, by using the bifurcation diagram study what happens when α and β are negative quantities.

Further reading

For additional information on the topics of the chapter, the following references may be consulted: [17], [22], [50], [54], [70], [84].

6

Cellular nonlinear networks

CONTENTS

Proceeding along the scheme of Figure 1.3, in this chapter complex systems modeled by many variables are taken into account. Let us consider the simplest nonlinear circuit we can conceive: a first-order circuit with the simplest nonlinearity, i.e., the saturation. Such simple system can be realized by means of an electronic analog circuit with one capacitor, one resistor, an active component, which can be a current or voltage generator, and voltage controlled current generators. By using only one circuit of this type it is difficult to understand what we can realize!

On the contrary, if we consider many of them, each locally connected with the others, that is, if we merge a lot of simple nonlinear electronic circuits, an emergent behavior could arise. This was the original idea of Leon O. Chua who in 1988 invented the so-called cellular neural networks (CNNs), then re-named cellular nonlinear networks and today cellular nanoscale networks.

CNNs are considered the paradigm of complexity in electronic circuit theory. They emphasize the role of cooperative behavior in the field of analog electronic circuits and from a computational point of view they represent an example of intense parallel processing. Despite the simplicity of the single cell, when it works together with other units of an array, high computational performance may be reached. Moreover, CNNs represented a scientific and technological revolution in the area of nonlinear circuits. In the modified universal complexity diagram the CNN circuits assume a dominant place.

The paradigm of CNN architecture opened a new frontier both in the circuit theory and, more in general, in the entire scientific community. The concept of synergetics proposed by Haken [39], who identified the properties of self-organization in non-equilibrium systems, is based on the fact that, from the nonlinear interaction of many subsystems, self-organization may macroscopically emerge. The CNN paradigm perfectly fits this concept introducing a real and tangible instance of it.

After the original paper by Chua and Yang in 1988 [21], different scientific communities dedicated their efforts on CNN studies, regarding both theoretical aspects and applications in different fields. CNN studies stimulated the improvement of technologies and VLSI design. A lot of books on CNNs have been published and the topic is still a fundamental one in various conferences and scientific societies. Since the beginning, every two years a conference dedicated to CNN studies has been regularly organized. Moreover, the CNN universal chip represents a milestone in information technology: it is a fully programmable 128×128, $1\mu m$ CMOS cell processor, where the main features are the digital interface of the analog core interfaced with an array of photo-sensors for direct optical image and video processing.

Even if high standard technologies lead to very sophisticated and fast devices (about 10^{15} operations per second), the absolutely simple and reliable concept of CNN allows scientists, researchers, and practitioners to realize parallel analog architectures in their labs with low-cost components. This is the additional power of CNNs: each of us, each student can build his CNN with few components in his home lab! Our efforts in this chapter will be devoted to giving self-contained information and material to do this.

6.1   CNN: basic notations

In its standard definition a CNN architecture consists of a rectangular array of $M \times N$ identical cells $C(i,j)$ where (i,j) with $i = 1, 2, \ldots, M$ and $j = 1, 2, \ldots, N$ indicates the spatial coordinates of the single cell, i indexing the rows and j the columns.

A schematized representation of a CNN is reported in Figure 6.1 in which each cell is locally linked with the others within a connection radius equal to

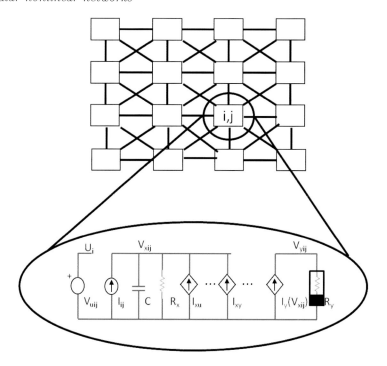

FIGURE 6.1
Schematic representation of a cellular nonlinear network. Each cell $C(i,j)$ is realized through the nonlinear circuit shown in the inset.

one. This means that each cell communicates locally with the nearest ones. In a general architecture a radius greater than one can also be considered. Indeed, the local connectivity permits the real-time propagation of the system variables to all cells. Information propagates similarly to a wave inside the network. Each cell, which is a continuous-time dynamical system (in the original CNN formalization it is a first-order system), is characterized by the following signals:

- the input signal $V_{u_{ij}}$;

- the state variable signal $V_{x_{ij}}$;

- the output signal $V_{y_{ij}}$.

The elementary electrical circuit of a single cell is reported in the inset of Figure 6.1. It is a first-order dynamical circuit in which the state variable is the voltage across capacitor C and the output nonlinearity is realized through the nonlinear resistor R_y whose current-voltage characteristic implements the saturation function:

$$V_{y_{ij}} = \frac{1}{2}(|V_{x_{ij}} + 1| - |V_{x_{ij}} - 1|) = f(V_{x_{ij}}). \qquad (6.1)$$

As concerns the other elements in the considered circuit, they are voltage-controlled current generators and, in particular:

- $I_{xu}(i, j; k, l) = B(i, j; k, l)V_{u_{kl}}$ is the current generated by the generator controlled by the input voltage $V_{u_{kl}}$ where $B(i, j; k, l)$ is a space dependent mathematical operator. Generically, it depends both on i, j and on the contour of the single cell;

- $I_{xy}(i, j; k, l) = A(i, j; k, l)V_{y_{kl}}$ is the current generated by the generator controlled by the output voltage $V_{y_{kl}}$ where $A(i, j; k, l)$ is a space dependent mathematical operator. Generically, it depends both on i, j, and on the contour of the single cell;

- I_{ij} is a constant current generator.

Given these definitions, and referring to the circuit in Figure 6.1, we apply the current Kirchoff law to the generic cell $C(i, j)$ to get:

$$C\frac{dV_{x_{ij}}}{dt} = -\frac{1}{R_x}V_{x_{ij}} + \sum_{C(k,l)\in N_r(i,j)} (A(i, j; k, l)V_{y_{kl}} + B(i, j; k, l)V_{u_{kl}}) + I_{ij}$$

$$(6.2)$$

with $1 \le i \le M$ and $1 \le j \le N$, while $N_r(i, j)$ is the r-neighborhood of cell $C(i, j)$ defined as

$$N_r(i, j) = \{C(k, l)| \max |k - i|, |l - j| \le r, 1 \le k \le M \ \& \ 1 \le l \le N\} \quad (6.3)$$

The operators A and B are called feedback and control template, respectively. Altogether A and B are referred to as the *cloning templates*. In practical applications space invariant templates and a 3×3 neighborhood (sphere of influence $r = 1$) are usually considered.

Example 6.1 ――――――――――――――――――――――――――――――――――――

Compute the following term:

$$\sum_{C(k,l)\in N_r(i,j)} A(i, j; k, l)Y_{kl}$$

where

$$A = \begin{bmatrix} a_{-1,-1} & a_{-1,0} & a_{-1,1} \\ a_{0,-1} & a_{0,0} & a_{0,1} \\ a_{1,-1} & a_{1,0} & a_{1,1} \end{bmatrix}$$

is a space invariant template and

$$Y_{ij} = \begin{bmatrix} y_{i-1,j-1} & y_{i-1,j} & y_{i-1,j+1} \\ y_{i,j-1} & y_{i,j} & y_{i,j+1} \\ y_{i+1,j-1} & y_{i+1,j} & y_{i+1,j+1} \end{bmatrix}$$

Solution. We make the operation $A * Y_{ij}$, where the symbol $*$ indicates the summation of dot product, i.e., the convolution operator, and we obtain the desired term:

$$\sum_{C(k,l) \in N_r(i,j)} A(i,j;k,l)Y_{kl} = A * Y_{ij} =$$
$$= a_{-1,-1}y_{i-1,j-1} + a_{-1,0}y_{i-1,j} + a_{-1,1}y_{i-1,j+1} + a_{0,-1}y_{i,j-1} + \quad (6.4)$$
$$+ a_{0,0}y_{i,j} + a_{0,1}y_{i,j+1} + a_{1,-1}y_{i+1,j-1} + a_{1,0}y_{i+1,j} + a_{1,1}y_{i+1,j+1}$$

We adopt the following notation:

- boundary cells: those cells belonging to the first and last columns and rows;

- corner cells: those cells located at the corners of the CNN.

The choice of appropriate boundary conditions is a fundamental point in problems of partial differential equations. In fact, the CNN represents a discrete-space, continuous-time system. Boundary conditions are applied considering a frame of additional cells around the $M \times N$ CNN, as represented in Figure 6.2 where the light gray cells compose the frame of a 5×5 CNN. The state variables of additional cells are set according to the following possible choices:

- fixed, or Dirichlet, boundary conditions: the additional cells provide an output fixed to a constant value;

- zero-flux, or Neumann, boundary conditions: the additional cells are constantly updated so that each of them provides an output which is equal to that of the proximal cell of the actual boundary;

- periodic, or toroidal, boundary conditions: the additional cells replicate the output of the actual boundary at the opposite side of the CNN.

By using the convolution operator and normalizing with respect to the time constant $R_x C$, the equation of each CNN cell becomes:

$$\frac{dx_{ij}}{dt} = -x_{ij} + A * Y_{ij} + B * U_{ij} + I \qquad (6.5)$$

with $1 \leq i \leq M$ and $1 \leq j \leq N$.

The model introduced with Equation (6.5) describes the dynamical behavior of a generic $M \times N$ CNN whose cells can be considered as placed over a two-dimensional plane. This model is often indicated as *single-layer* CNN. Thus, in the single-layer model, each cell has a single state variable. Furthermore, it is worth noting that in Equation (6.5) the cloning templates are considered as being space invariant. Due to the fact that the main applications of CNNs make use of this assumption, in the following sections cloning templates are considered space invariant. The model can be generalized to multiple state variables considering several stacked single-layer CNNs. A multi-layer CNN

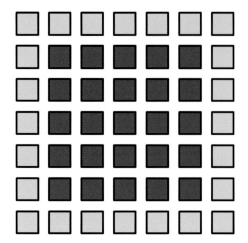

FIGURE 6.2
Additional cells for a 5×5 CNN representing boundary conditions.

(or ML-CNN) can also be viewed as a single-layer CNN in which each cell has several state variables. Moreover, the interaction between the state variables of the same cell can be complete, while the cell to cell interaction remains local (restricted to the r-neighborhood).

A formal model for a ML-CNN is the following:

$$\frac{d\mathbf{x}_{ij}}{dt} = -\mathbf{x}_{ij} + \mathbf{A} * \mathbf{Y}_{ij} + \mathbf{B} * \mathbf{U}_{ij} + \mathbf{I} \tag{6.6}$$

with $1 \leq i \leq M$ and $1 \leq j \leq N$ and where:

$$
\begin{aligned}
\mathbf{A} &= \begin{bmatrix} \mathrm{A}_{11} & \cdots & \mathrm{A}_{1m} \\ \vdots & \ddots & \vdots \\ \mathrm{A}_{m1} & \cdots & \mathrm{A}_{mm} \end{bmatrix} \\
\mathbf{B} &= \begin{bmatrix} \mathrm{B}_{11} & \cdots & \mathrm{B}_{1m} \\ \vdots & \ddots & \vdots \\ \mathrm{B}_{m1} & \cdots & \mathrm{B}_{mm} \end{bmatrix} \\
\mathbf{I} &= \begin{bmatrix} I_1, \ldots, I_m \end{bmatrix}^T \\
\mathbf{x}_{ij} &= \begin{bmatrix} x_{1,ij}, \ldots, x_{m,ij} \end{bmatrix}^T \\
\mathbf{y}_{ij} &= \begin{bmatrix} y_{1,ij}, \ldots, y_{m,ij} \end{bmatrix}^T \\
\mathbf{u}_{ij} &= \begin{bmatrix} u_{1,ij}, \ldots, u_{m,ij} \end{bmatrix}^T
\end{aligned} \tag{6.7}
$$

where m denotes the number of the state variables in the multi-layer cell circuits or, in other words, the number of layers. We remark that in this case the convolution operator $*$ has to be intended as a matrix multiplication in which the convolution is applied to each entry.

Example 6.2

Write a CNN model with one cell and constant templates.
Solution. It is:

$$\dot{x}_1 = -\frac{1}{R_x C} x_1 + ay_1(t) + bu_1(t) + I \qquad (6.8)$$

where $y_1(t) = \frac{1}{2}(|x_1 + 1| - |x_1 - 1|)$ and a, b, and I scalar, real quantities.

Example 6.3

Write a CNN model with two cells assuming $\frac{1}{R_x C} = 1$ and without input signals.
Solution. The architecture of the CNN is a 1×2 CNN ($M = 1$, $N = 2$). The model consists of two state equations as follows:

$$\begin{aligned} \dot{x}_1 &= -x_1 + a_{11}y_1(t) + a_{12}y_2(t) \\ \dot{x}_2 &= -x_2 + a_{21}y_1(t) + a_{22}y_2(t) \end{aligned} \qquad (6.9)$$

where $y_1(t) = \frac{1}{2}(|x_1 + 1| - |x_1 - 1|)$, $y_2(t) = \frac{1}{2}(|x_2 + 1| - |x_2 - 1|)$ and a_{11}, a_{12}, a_{21} and a_{22} are scalar, real quantities.
We remark that this model may represent both a two-cell CNN and a two-layer CNN, each of which with one cell.

Example 6.4

Let us now consider the following model:

$$\begin{aligned} \dot{x}_1 &= -x_1 + \alpha y_1 + \beta y_2 \\ \dot{x}_2 &= -x_2 + \gamma y_1 + \alpha y_2 \end{aligned} \qquad (6.10)$$

Indeed this two-cell CNN can be rewritten in the standard form assuming boundary conditions fixed to null values. In fact assuming:

$$A = \begin{bmatrix} 0 & 0 & 0 \\ \gamma & \alpha & \beta \\ 0 & 0 & 0 \end{bmatrix} \qquad Y = \begin{bmatrix} 0 & y_1 & y_2 \\ y_1 & y_2 & 0 \end{bmatrix} \qquad (6.11)$$

we have:

$$A * Y = \begin{bmatrix} \alpha y_1 + \beta y_2 \\ \gamma y_1 + \alpha y_2 \end{bmatrix} \qquad (6.12)$$

and the standard state space form is obtained.

6.2 CNNs: main aspects

The CNN scheme represents an analog system made by a huge number of differential equations. Let us consider, for example, a 128×128 CNN, its dynamics includes more than 13000 state variables. The corresponding nonlinear dynamical system is solved in real-time in a few milliseconds. This means that the steady-state stable equilibrium point, if it exists, is reached in a short time, with high speed. Note also that we are considering nonlinear systems with a great number of equilibrium points so that the end point depends on the initial state and on the input signals. The previous considerations lead us to consider CNNs as very powerful analog circuits able to solve nonlinear partial differential equations. Moreover, they can constitute associative memories.

Indeed, the power of CNNs derives from their architecture and the nonlinear behavior of its dynamics. If we consider a CNN and an r-neighborhood with $r = 1$, the dynamics of the system depends on the 9 coefficients of the template A and on the 9 coefficients of the template B. Including the bias constant current I, we have a total of 19 parameters. Hence, by considering space invariant cloning templates, we get a powerful system with more than 10000 equations depending only on 19 parameters!

By using different cloning templates A and B we can program the CNN in order to obtain different operations. By choosing the cloning templates we build a dynamical system whose steady-state solution gives the desired processed information starting from a matrix of initial conditions and of input signals. The key point in the development of CNN systems is to create a suitable family of cloning templates that solves some basic operations. Then, executing one after the other such simple tasks, complex algorithms for image processing can be implemented.

Four aspects of CNNs must be remarked:

- they have been conceived as nonlinear image processing architecture;

- the images to process are used to initialize inputs and/or state variables, thus the size of the CNN is equal to the size (in pixels) of the considered image;

- gray-scale images must be normalized in the range $[-1, 1]$, where -1 maps white, 1 maps black, and 0 the average gray;

- from their intrinsic structure CNNs are solvers of partial differential equations, that are discrete in space and continuous in time.

In the next section we will discuss the choice of cloning templates to perform elementary image processing operations.

6.3   Cloning templates and features of CNNs

It is well known that image processing in real-time is a difficult computational task and that classical segmentation and interpolation operations are conditioned by the processor used to perform them. For their high performance in carrying out parallel processing and their high speed of execution, CNNs are a valid alternative method to the classical processing techniques. The cloning templates A and B to perform standard image processing have been derived heuristically taking into account some guidelines that have been successfully used in spatial digital filters. Furthermore, libraries of standard templates are available [44].

Let us consider the simple CNN model of Equations (6.5) and a two-dimensional array of data, such as a gray-scale digital image. Suppose we set these data as initial values for the cell state variables (normalized in the $[-1, 1]$ range). This is the image that must be processed. Here we consider no input, i.e., $B(i, j; k, l) = 0$ or $u_{ij} = 0$. As the CNN dynamical system evolves, the state is changed according to the cloning templates. When the steady state is reached, the state values (or the output) represent the processed array of data, so the processed image. When $B(i, j; k, l) \neq 0$ another array of data is available for a new processing (it can be used as a second operand). It is worth noting that this is a real-time processing where the cloning templates are dual to computer instructions, while initial state and input are dual to operands. Following this principle, CNNs are applied in the field of image processing.

Moreover, more complex transformations can be achieved by subsequently processing different templates, exactly like a datum can be processed by the various instructions of a computer program. This duality among the templates and the computer instructions is the basis of the above cited dual computing and led to the introduction of the CNN Universal Machine (CNN-UM).

6.3.1 A simple software implementation of a CNN

Let us now discuss the possible software implementation of the CNN architecture. The dynamical process defined by a CNN can be easily implemented through the numerical integration of the Equations (6.5). In this section, a simple MATLAB® procedure to implement the CNN dynamics is described. The main commands to simulate a $M \times N$ CNN are reported in the following:

```
function [X,Y,Xd,Yd]=cnn(X1,U1,A,B,I,boundCondType,dt,maxstep)
[N1,M1]=size(X1);
X1=normalizeCNN(X1);
U1=normalizeCNN(U1);
N=N1+2;
M=M1+2;
X=zeros(N,M,maxstep);
Y=zeros(N,M,maxstep);
U=zeros(N,M);
U(2:end-1,2:end-1)=U1;
X(2:end-1,2:end-1,1)=X1;
Y(:,:,1)=0.5*(abs(X(:,:,1)+1)-abs(X(:,:,1)-1));
for t=2:maxstep
    for i=2:(N1+1)
        for j=2:(M1+1)
            AY=0;
            BU=0;
            for k=[-1 0 1]
                for l=[-1 0 1]
                    AY=AY+A(k+2,l+2)*Y(i+k,j+l,t-1);
                    BU=BU+B(k+2,l+2)*U(i+k,j+l);
                end
            end
            X(i,j,t)=X(i,j,t-1)+dt*(-X(i,j,t-1)+AY+BU+I);
            Y(i,j,t)=0.5*(abs(X(i,j,t)+1)-abs(X(i,j,t)-1));
        end
    end
end
```

```
    if boundCondType~=1
        [Xt,Yt]=Boundary(squeeze(X(:,:,t)),squeeze(Y(:,:,t)),type);
        X(:,:,t)=Xt;
        Y(:,:,t)=Yt;
    end
end
Xd=denormalizeCNN(X(:,:,end));
Yd=denormalizeCNN(Y(:,:,end));
Xd=Xd(2:end-1,2:end-1);
Yd=Yd(2:end-1,2:end-1);
end
```

The cnn function needs eight inputs and provides four outputs. Inputs are:

- an $M \times N$ matrix X1 containing the initial states $x_{ij}(0)$;

- an $M \times N$ matrix U1 containing the inputs u_{ij};

- the cloning templates A, B, and I;

- an integer boundCondType representing the desired boundary conditions, namely 1 for fixed boundary conditions, 2 for zero-flux boundary conditions, and 3 for periodic boundary conditions;

- the integration step dt;

- the total number of integration steps maxstep.

The outputs of the functions are:

- an $M \times N \times$maxstep matrix X containing the whole evolution of the states $x_{ij}(t)$;

- an $M \times N \times$maxstep matrix Y containing the whole evolution of the outputs $y_{ij}(t)$;

- an $M \times N$ matrix Xd representing the denormalized final state;

- an $M \times N$ matrix Yd representing the denormalized final output.

Due to the fact that one of the main purposes of CNNs is nonlinear image processing, the function is written so that the input gray-scale image is normalized in the range $[-1, 1]$ according to the CNN paradigm, in such a way that the white color corresponds to -1, while the black color is mapped to 1. The normalizeCNN function is the following:

```
function Xn=normalizeCNN(X)
    Xn=-(2*(X./255)-1);
end
```

where we assumed that the image is given in gray-scale and each pixel is in the interval $[0, 255]$. To denormalize the final output of the CNN restoring the standard gray-scale it suffices to apply the inverse of the normalizing formula as implemented by the following MATLAB® function:

```
function Xn=denormalizeCNN(X)
    Xn=-(X-1).*255./2;
end
```

The core of the function **cnn** consists in the two nested **for** cycles spanning the $N \times M$ CNN and integrating with the Euler method the dynamical equations of each cell. The coupling is weighted through the cloning templates and cells are synchronously updated, i.e., first all derivatives are calculated based on previous values and, then, the values of all cells are updated. As concerns boundary conditions, in the case of fixed values they are set to zero while in the other two cases the **Boundary** function, which is implemented as follows, is called:

```
function [Xt,Yt]=Boundary(X,Y,type)
if (type==2) %%Zero-flux Boundary conditions
    Xt=X;
    Xt(:,1)=X(:,2);
    Xt(:,end)=X(:,end-1);
    Xt(1,:)=X(2,:);
    Xt(end,:)=X(end-1,:);
    Yt=Y;
    Yt(:,1)=Y(:,2);
    Yt(:,end)=Y(:,end-1);
    Yt(1,:)=Y(2,:);
    Yt(end,:)=Y(end-1,:);
else %%Periodic Boundary conditions
    Xt=X;
    Xt(:,1)=X(:,end-1);
    Xt(:,end)=X(:,2);
    Xt(1,:)=X(end-1,:);
    Xt(end,:)=X(2,:);
    Yt=Y;
    Yt(:,1)=Y(:,end-1);
    Yt(:,end)=Y(:,2);
    Yt(1,:)=Y(end-1,:);
    Yt(end,:)=Y(2,:);
end
end
```

In the following, we give some examples of image processing tasks performed by using the **cnn** function.

Example 6.5

Edge extraction of an image.
Let us consider the problem of extracting the edges of the input image shown in Figure 6.3(a). The aim is to extract the edges of each shape where black pixels having at least one white neighbor are considered part of the edge of the object. This task is solved with the following cloning templates:

$$A = \begin{bmatrix} 0 & 0 & 0 \\ 0 & 1 & 0 \\ 0 & 0 & 0 \end{bmatrix}; \quad B = \begin{bmatrix} -1 & -1 & -1 \\ -1 & 8 & -1 \\ -1 & -1 & -1 \end{bmatrix}; \quad I = -1 \qquad (6.13)$$

The input U of the CNN is the considered image, while the initial states can be set arbitrarily. Boundary conditions fixed to zero are used. The input image can be loaded in the MATLAB® workspace using the following command:

```
X1=double(imread('image.bmp'));
```

The same command line will be used in all the following examples to load images to be processed. The simulation can now be run by using the following command line:

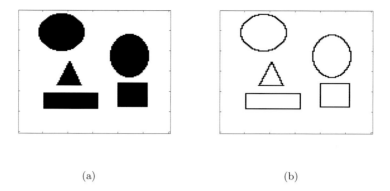

(a)　　　　　　　　　　　　　　　　　　　　(b)

FIGURE 6.3

Edge extraction: (a) gray-scale image used as input in Examples 6.5 and 6.6, (b) output of the CNN with templates (6.13).

```
[X,Y,Xd,Yd]=cnn(zeros(121),X1,[0 0 0;0 1 0;0 0 0],
    [-1 -1 -1;-1 8 -1;-1 -1 -1 ],-1,1,0.1,100);
```

After 100 time steps, the output of the CNN is reported in Figure 6.3(b) where the edges of the five shapes are clearly retrieved.

Example 6.6

Corner detection.
For this task, that aims to find the convex corners of the objects, defined as the black pixels having at least 5 white neighbors, we consider again the picture shown in Figure 6.3(a) and use the following set of cloning templates:

$$A = \begin{bmatrix} 0 & 0 & 0 \\ 0 & 1 & 0 \\ 0 & 0 & 0 \end{bmatrix}; \quad B = \begin{bmatrix} -1 & -1 & -1 \\ -1 & 4 & -1 \\ -1 & -1 & -1 \end{bmatrix}; \quad I = -5 \qquad (6.14)$$

The input U is the image, while also in this case the initial states can be set arbitrarily. Boundary conditions fixed to zero are used. The simulation is started using the following command line:

```
[X,Y,Xd,Yd]=cnn(zeros(121),X1,[0 0 0;0 1 0;0 0 0],
    [-1 -1 -1;-1 4 -1;-1 -1 -1 ],-5,1,0.1,100);
```

After 100 time steps, the output of the CNN is that reported in Figure 6.4 where the corners of the triangular and rectangular shapes are clearly retrieved.

Example 6.7

Noise removal.
In order to eliminate Gaussian noise from the picture shown in Figure 6.5(a) the following templates are used:

$$A = \begin{bmatrix} 0 & 1 & 0 \\ 1 & 2 & 1 \\ 0 & 1 & 0 \end{bmatrix}; \quad B = \begin{bmatrix} 0 & 0 & 0 \\ 0 & 0 & 0 \\ 0 & 0 & 0 \end{bmatrix}; \quad I = 0 \qquad (6.15)$$

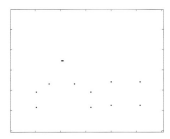

FIGURE 6.4

Corner detection of shapes in an image: output of the CNN with templates (6.14).

Note that only the feedback template is used. The image to process is used both as initial state of the CNN and as input. Zero-flux boundary conditions are considered. Note also that the template A is structured so that it reinforces the effect of reducing the pixel noise, by using a feedback signal that weights in a positive way the neighboring cells. Simulation can be run using the command:

```
[X,Y,Xd,Yd]=cnn(X1,X1,[0 0.5 0;0.5 1 0.5;0 0.5 0],
    [0 0 0 ;0 0 0 ;0 0 0],0,2,0.1,100);
```

After 100 time steps, the output of the CNN is that reported in Figure 6.5(b) where noise has been removed from the original image.

Example 6.8 _____

Deleting diagonal lines.
Let us consider the picture shown in Figure 6.6(a), used to illustrate an example where the aim is to delete only the diagonal lines, preserving both the vertical and the horizontal ones. For this purpose we use the following cloning templates:

$$A = \begin{bmatrix} 0 & 0 & 0 \\ 0 & 1 & 0 \\ 0 & 0 & 0 \end{bmatrix} \quad B = \begin{bmatrix} -1 & 0 & -1 \\ 0 & 1 & 0 \\ -1 & 0 & -1 \end{bmatrix} \quad I = -4 \qquad (6.16)$$

The input of the CNN is the image of interest, while initial states are set to zero. Boundary conditions fixed to zero are used. The templates are designed in order to delete diagonal lines with a single pixel width. The CNN is simulated with the command:

```
[X,Y,Xd,Yd]=cnn(zeros(48),X1,[0 0 0;0 1 0;0 0 0],
    [-1 0 -1 ;0 1 0 ;-1 0 -1],-4,1,0.1,100);
```

After 100 time steps, the diagonal lines observable in the original picture are correctly deleted as reported in Figure 6.6(b).

Example 6.9 _____

Deleting vertical lines.
A task similar to Example 6.8 is considered here, that is deleting only vertical lines. To show the capability of a CNN to perform such an operation we consider again the image of Figure 6.6(a) and apply the following cloning templates:

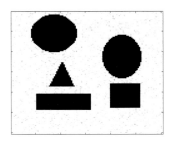

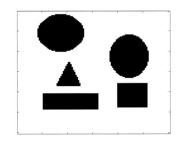

(a)   (b)

FIGURE 6.5
Gaussian noise removal: (a) input image on which a Gaussian noise has been applied, (b) output of the CNN with templates (6.15).

$$A = \begin{bmatrix} 0 & 0 & 0 \\ 0 & 1 & 0 \\ 0 & 0 & 0 \end{bmatrix}; \quad B = \begin{bmatrix} 0 & -1 & 0 \\ 0 & 1 & 0 \\ 0 & -1 & 0 \end{bmatrix}; \quad I = -2 \qquad (6.17)$$

The image is used as input, the initial states are set to zero, boundary conditions fixed to zero are used. The set of templates deletes black pixels having two black neighbors in the vertical direction. The command to run the CNN simulation is:

```
[X,Y,Xd,Yd]=cnn(zeros(48),X1,[0 0 0;0 1 0;0 0 0],
    [0 -1 0 ;0 1 0 ;0 -1 0],-2,1,0.1,100);
```

After 100 time steps, the vertical lines of the input picture are removed as shown in Figure 6.7.

Example 6.10 ⎯⎯⎯⎯⎯⎯⎯⎯⎯⎯⎯⎯⎯⎯⎯⎯⎯⎯⎯⎯⎯⎯⎯⎯⎯⎯⎯⎯⎯⎯⎯⎯⎯⎯⎯⎯

CNN Universal Machine for noise removal and corner extraction.
In a CNN different image processing tasks can be made sequentially to obtain a more complex image processing operation. Here we show an example of such operations performed by the CNN Universal Machine. Two elementary image processing tasks are performed sequentially by using the output of the first operation as the input of the second. Consider again the noisy image shown in Figure 6.5(a): we aim at removing noise and then extract the corners of the shapes.
The two CNN processes are simulated by means of the following commands:

```
[X_noise,Y_noise,Xd_noise,Yd_noise]=cnn(X1,X1,
    [0 0.5 0;0.5 1 0.5;0 0.5 0],[0 0 0 ;0 0 0 ;0 0 0],0,2,0.1,100);
[X_corner,Y_corner,Xd_corner,Yd_corner]=cnn(zeros(121),Yd_noise,
    [0 0 0;0 1 0;0 0 0],[-1 -1 -1;-1 4 -1;-1 -1 -1 ],-5,1,0.1,100);
```

The output of the second processing step is that reported in Figure 6.4.

To give an idea of the order of magnitude of the speed required by a CNN Universal Machine, we mention that the typical speed has been estimated in 20 trillion operations per second. Indeed, in a classical digital computer in order

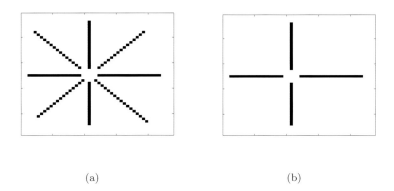

(a)   (b)

FIGURE 6.6
Deletion of diagonal lines from an image: (a) input image with diagonal, vertical, and horizontal lines, (b) output of the CNN with templates (6.16).

to compute the spatiotemporal dynamics for a structure with 10000 cells two million equivalent digital operations for single unit time must be performed while in the analog CNN chip the final integration time (i.e., the time at which the dynamical system reaches the steady state) is obtained in about $100ns$. This is a representative value which does not take into consideration the technology of advanced CNN chips realized in $5\mu m$.

6.3.2 Choice of the templates

One of the main problems of CNN is how to choose the templates that permit to obtain a desired processing task. This is known as the learning and design problem. Approaching the choice of the cloning templates as a design problem is equivalent to building a dynamical system with the constraint of the CNN model and choosing the templates in such a way that the dynamics of the designed system evolves in the desired manner. Practically, we have seen that the design problem is solved when a desired task could be translated into a set of local dynamical rules. The dual learning problem is considered when the templates are obtained by using learning algorithms aimed at finding the coefficient values so that pairs of inputs and outputs are matched.

Another interesting approach to select the cloning templates is the direct mapping between a partial differential equation and a CNN model. If an equivalence between the two models can be established, then the templates are directly derived.

Example 6.11 _____

Solve the heat diffusion equation with a CNN.

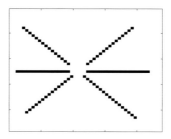

FIGURE 6.7
Deletion of vertical lines from an image: output of the CNN with templates
(6.17).

Solution. This example is a milestone in the design of CNN systems able to solve
differential equations with Laplacian operator. Let us consider the following equation:

$$\frac{\partial P}{\partial t} = \frac{k^2}{c^2}\nabla^2 P = \frac{k^2}{c^2}\left(\frac{\partial^2}{\partial x^2} + \frac{\partial^2}{\partial y^2}\right)P \tag{6.18}$$

where P is the temperature over a surface, k the thermal conductivity, and c the
heat capacity.
In order to map Equation (6.18) onto a CNN system, a spatial dicretization has
to be introduced. Therefore we indicate with P_{ij} the temperature of cell (i,j) and
the Laplacian $\nabla^2 P$ can be spatially discretized by using the following formula:

$$\nabla^2 P \simeq P_{i-1,j} + P_{i,j-1} + P_{i,j+1} + P_{i+1,j} - 4P_{i,j} \tag{6.19}$$

The discretized Laplacian works as a spatial operator processing the neighboring
cells in accordance with their indices

$$\begin{bmatrix} & i-1,j & \\ i,j-1 & i,j & i,j+1 \\ & i+1,j & \end{bmatrix}$$

Therefore, the mapping procedure between the discretized Laplacian model and
the standard CNN model lead to the following structure of the cloning templates:

$$A = \begin{bmatrix} 0 & \alpha & 0 \\ \alpha & -4\alpha & \alpha \\ 0 & \alpha & 0 \end{bmatrix}; \quad B = \begin{bmatrix} 0 & 0 & 0 \\ 0 & 0 & 0 \\ 0 & 0 & 0 \end{bmatrix}; \quad I = 0 \tag{6.20}$$

where $\alpha = \frac{k^2}{c^2}$ is the thermal diffusivity. The CNN system reads as:

$$\dot{P}_{ij} = A * Y_{ij} \tag{6.21}$$

where $Y_{ij} = \begin{bmatrix} P_{i-1,j-1} & P_{i-1,j} & P_{i-1,j+1} \\ P_{i,j-1} & P_{i,j} & P_{i,j+1} \\ P_{i+1,j-1} & P_{i+1,j} & P_{i+1,j+1} \end{bmatrix}$.

This example emphasizes the outstanding use of CNN systems as partial differential
equation solvers.

6.4   The CNN as a generator of nonlinear dynamics

Section 6.3 introduced several methods to generate CNNs. We now go further and investigate the question: since a CNN is a dynamical nonlinear system, can an arbitrary dynamical system be emulated by a CNN? To state in different terms, given a dynamical system can we realize its dynamics by using the CNN model and so can we fix A and B in such a way that the CNN model has a dynamics equivalent to the given system?

Let us consider the elementary CNN cell, that we can view as a simple brick. Can we realize a general dynamical system by coupling more than one cell? And, therefore, from the coupling of dynamical cells, does the CNN generate a more complex dynamics?

Given that a CNN is a programmable and extremely flexible circuit, it is natural to ask the question: what kind of nonlinear dynamics can be obtained from a CNN? What would be the consequences of a positive answer to this question? First of all, we have a theoretical consequence: the CNN would represent a general model for the class of circuits for which it reproduces the dynamics. In other words, the cell would be the primitive of a wide class of dynamical circuits. To provide the guidelines to establish what we called "procedure for the design of cells" a further definition of a new class (more complete) of CNN must be given.

The state-controlled cellular neural network (SC-CNN) is a generalization of the standard CNN architecture obtained by introducing a further template, namely $\hat{A}$ indicated as the linear feedback template, which encompasses in Equation (6.5) a term related to the state variables of the cells. The following model is referred to as a one-layer SC-CNN, using the convolution formalism described above:

$$\dot{x}_{ij} = -x_{ij} + A * Y_{ij} + \hat{A} * X_{ij} + B * U_{ij} + I_{ij} \qquad (6.22)$$

with $1 \leq i \leq M$ and $1 \leq j \leq N$. The $\hat{A}$ operator is also said to be the *state template*. The output nonlinearity is the standard PWL one, as defined above.

The procedure to realize arbitrary dynamical circuits as SC-CNNs is based on the following steps:

- define the discrete component realization of SC-CNN cells;

- generate nonlinear dynamics by using SC-CNN by using a number of cells equal to the system order;

- define a strategy to realize nonlinearity different from the canonical PWL one;

- implement the nonlinear systems with circuits based on the SC-CNN cells.

These considerations lead to the conclusion that nonlinear systems with complex dynamics can be realized by CNN cells. So, although the cells are simple, that is, their uncoupled dynamics displays a trivial behavior, coupling them, as the CNN philosophy suggests, leads to the emergence of the complex dynamics.

6.4.1 Discrete component realization of SC-CNN cells

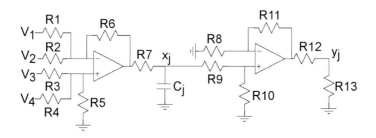

FIGURE 6.8
Circuit realization of an SC-CNN cell.

Let us consider the scheme reported in Figure 6.8. It represents the circuit realization, based on standard components, of an SC-CNN cell. Firstly, the *gain rule* for the two operational amplifiers, must be satisfied, i.e., the sum of conductances at the non-inverting terminal must be equal to the sum of conductances at the inverting terminal, so that the following relationships must hold:

$$\frac{1}{R_1} + \frac{1}{R_2} + \frac{1}{R_6} = \frac{1}{R_3} + \frac{1}{R_4} + \frac{1}{R_5}$$

$$\frac{1}{R_8} + \frac{1}{R_{11}} = \frac{1}{R_9} + \frac{1}{R_{10}}$$

(6.23)

Under this assumption, the output of the operational amplifiers is

$$v_o = \sum_i A_i v_i \tag{6.24}$$

where $A_i = \frac{R_f}{R_i}$. The state equation of the circuit can be thus written as:

$$C_j \dot{x}_j = -\frac{1}{R_7} x_j + \frac{R_6}{R_7 R_1} V_1 + \frac{R_6}{R_7 R_2} V_2 - \frac{R_6}{R_7 R_3} V_3 - \frac{R_6}{R_7 R_4} V_4 \tag{6.25}$$

while the nonlinearity is implemented exploiting the saturation of the operational amplifiers as

$$y_j = \frac{R_{13}}{R_{12} + R_{13}} \frac{R_{11}}{R_9} (|x_j + 1| - |x_j - 1|) \tag{6.26}$$

A basic cell for each state variable of the dynamical system to implement has to be considered, where the state variable is associated to the voltage across capacitor C_j. The values of the circuit components are chosen to match the two dynamics, that is, to match Equation (6.25) with each state equation of the dynamical system to implement. In the following, some examples illustrating the design procedure are reported.

6.4.2 Chua's circuit dynamics generated by the SC-CNN cells

Let us discuss the SC-CNN based implementation of the Chua's circuit dynamics, as modeled by the dynamical equations (5.5) with PWL nonlinearity. For the implementation we fix the parameters so that to obtain the double scroll Chua's attractor: $m_0 = -\frac{1}{7}$, $m_1 = \frac{2}{7}$, $\alpha = 9$, and $\beta = 14.286$.

The electronic circuit is designed using three cells of the type shown in Figure 6.8. The circuit is based on off-the-shelf resistors, capacitors and operational amplifiers. The complete scheme is reported in Figure 6.9 where, since only the x variable of the Chua's circuit drives the nonlinearity of the system, only the first cell, implementing that variable, presents the nonlinear output stage.

The circuit equations associated to the implementation of Chua's circuit are the following:

$$\begin{cases} C_1 R_6 \frac{dX}{d\tau} = -X + \frac{R_5}{R_3} Y + \frac{R_5}{R_2} h \\ C_2 R_{18} \frac{dY}{d\tau} = -Y + \frac{R_{17}}{R_{14}} X + \frac{R_{17}}{R_{15}} Z \\ C_3 R_{23} \frac{dZ}{d\tau} = -Z + \frac{R_{21}}{R_{20}} Z + \frac{R_{21}}{R_{19}} Y \end{cases} \tag{6.27}$$

where:

$$h = \frac{R_{12}}{R_{11} + R_{12}} \frac{R_9}{R_8} (|X + 1| - |X - 1|) \tag{6.28}$$

Matching Equations (6.27) with the mathematical model (5.5) makes possible the choice of the values of the components. At this stage a temporal rescaling allowing for faster observation of the dynamical behavior is introduced. In particular, we selected a rescaling factor $\kappa = \frac{1}{C_2 R_{18}} = \frac{1}{C_3 R_{23}} = 10000$. Furthermore, choosing R_6 as a variable resistor, the different dynamical behaviors shown by the Chua's circuit by varying the single bifurcation parameter α can be observed. In fact, the parameter is linked to the resistor value through the relationship $\alpha = \frac{R_5}{R_3} \frac{R_{18}}{R_6}$.

Finally, we summarize here the templates needed to implement the Chua's circuit following the SC-CNN architecture:

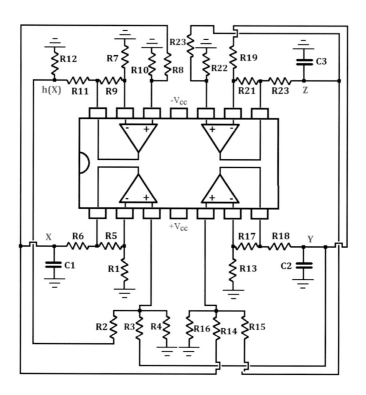

FIGURE 6.9
SC-CNN based circuit implementation of the Chua's equations (5.5). Components: $R_1 = 4k\Omega$, $R_2 = 13.3k\Omega$, $R_3 = 5.6k\Omega$, $R_4 = 20k\Omega$, $R_5 = 20k\Omega$, $R_6 = 380\Omega$ (potentiometer), $R_7 = 112k\Omega$, $R_8 = 112k\Omega$, $R_9 = 1M\Omega$, $R_{10} = 1M\Omega$, $R_{11} = 8.2k\Omega$, $R_{12} = 1k\Omega$, $R_{13} = 51.1k\Omega$, $R_{14} = 100k\Omega$, $R_{15} = 100k\Omega$, $R_{16} = 100k\Omega$, $R_{17} = 100k\Omega$, $R_{18} = 1k\Omega$, $R_{19} = 8.2k\Omega$, $R_{20} = 100k\Omega$, $R_{21} = 100k\Omega$, $R_{22} = 7.8k\Omega$, $R_{23} = 1k\Omega$, $C_1 = C_2 = C_3 = 100nF$, $V_{cc} = 9V$. Components with 1% tolerances are used.

$$A = \begin{bmatrix} \alpha(m_0 - m_1) & 0 & 0 \\ 0 & 0 & 0 \\ 0 & 0 & 0 \end{bmatrix}; \quad \hat{A} = \begin{bmatrix} 1 - \alpha m_1 & \alpha & 0 \\ 1 & 0 & 1 \\ 0 & -\beta & 1 \end{bmatrix};$$

$$B = \begin{bmatrix} 0 & 0 & 0 \\ 0 & 0 & 0 \\ 0 & 0 & 0 \end{bmatrix}; \quad I = \begin{bmatrix} 0 & 0 & 0 \end{bmatrix} \qquad (6.29)$$

6.4.3 A generalized cell for realizing any multivariable nonlinearities using PWL functions

The case of Chua's circuit can be directly mapped on the SC-CNN model due to the fact that the nonlinearity is a saturation such as that defined in the general model of CNNs. However, it is also possible to deal with more complex cases. This relies on two important assertions: (1) any nonlinear function can be approximated with a given precision combining a finite number of PWL functions [32], and (2) any PWL function can be obtained as the sum of a definite number of CNN standard PWL functions [8]. From the circuit point of view, it means that more output stages may be required but, since the standard PWL is the universal atomic function, all of them can be designed following the same schematic.

The Equations (6.27) refer to a three-layer SC-CNN architecture. More precisely, it is a SC-CNN with space invariant templates. However, the same dynamics can be implemented considering a planar 1×3 SC-CNN with space dependent templates. This aspect allows us to assess that *any* nonlinear dynamical system can be implemented as an SC-CNN, at most introducing a given approximation.

The result is encouraging for two reasons:

1. it allows us to implement complex dynamics by using elementary cells, so that the cell is the main building element of all dynamical systems;

2. following the proposed approach we observe that mapping is a good strategy to select templates.

The conclusion is that you can easily realize your own low cost CNN in your lab!

6.5 Reaction-diffusion CNN

Many nonlinear phenomena like solitons, autowaves, and spiral waves have been observed and studied by biologists, neurologists, and physicians in nonlinear active media [60]. The model that characterizes them is the classical reaction-diffusion equation:

$$\dot{\mathbf{x}} = F(\mathbf{x}, t) + D\nabla^2\mathbf{x} \qquad (6.30)$$

where $\mathbf{x} \in \mathbb{R}^n$ is the state vector, $F : \mathbb{R}^{n+1} \to \mathbb{R}^n$ represents the reactive term, and $D\nabla^2\mathbf{x}$ is the diffusive term in which is included the Laplacian $\nabla^2\mathbf{x}$. Therefore a reaction that is modeled by the nonlinear dynamical system

$$\dot{\mathbf{x}} = F(\mathbf{x}, t)$$

diffuses its effect in a medium thanks to the action of the diffusive term $D\nabla^2\mathbf{x}$. If the phenomenon occurs in active media there is the possibility that autowaves and active waves are generated. Reaction-diffusion equations can be solved mapping their dynamics on the CNN (or SC-CNN) architecture. We start by describing the model of a universal cell that could represent the local reaction of an electronic cell.

6.5.1 The reaction CNN cell

In order to derive the model for a cell suitable to describe reaction-diffusion equations let us start from the following examples.

Example 6.12 _____

Let us consider the model introduced in Example 6.4 where it is assumed $\gamma = -\beta$ with $\beta < 0$, and, in particular, $\alpha = 2$ and $\beta = -2$. The model can be written as:

$$\begin{aligned} \dot{x}_1 &= -x_1 + 2y_1 - 2y_2 \\ \dot{x}_2 &= -x_2 + 2y_1 + 2y_2 \end{aligned} \qquad (6.31)$$

with y_1 and y_2 defined following the CNN standard definition.
To study the dynamical behavior of this system, we numerically solve Equations (6.31) using the function twocells:

```
function dxdt=twocells(t,x,alpha,beta)

y1=0.5*(abs(x(1)+1)-abs(x(1)-1));
y2=0.5*(abs(x(2)+1)-abs(x(2)-1));

dxdt=[-x(1)+alpha*y1+beta*y2
    -x(2)-beta*y1+alpha*y2];
```

and integrating it with the command:

```
[T,Y]=ode45(@twocells,[0:0.001:500],[0.1 0.5],'',2,-2);
```

The dynamical behavior of the system is reported in Figure 6.10 illustrating the (periodic) temporal evolution of the two state variables and the trajectory (a limit cycle) in the phase plane.

Example 6.13 _____

Let us consider the following non-autonomous CNN model in standard form:

$$\begin{aligned} \dot{x}_1 &= -x_1 + 2y_1 - 1.2y_2 + 4.04\sin\left(\tfrac{\pi}{2}t\right) \\ \dot{x}_2 &= -x_2 + 1.2y_1 + 2y_2 \end{aligned} \qquad (6.32)$$

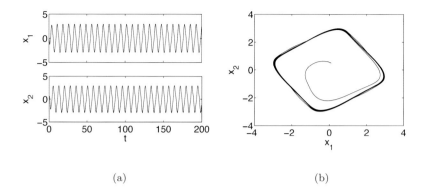

(a)   (b)

FIGURE 6.10
Dynamical behavior of the model (6.31): (a) temporal trend of the two state variables, and (b) trajectory of the system in the phase plane.

We aim at studying the dynamical behavior of this system by means of numerical integration of its dynamics. The function implementing the dynamical equations `twocellsNA` is:

```
function dxdt=twocellsNA(t,x,alpha,beta)

y1=0.5*(abs(x(1)+1)-abs(x(1)-1));
y2=0.5*(abs(x(2)+1)-abs(x(2)-1));

dxdt=[-x(1)+alpha*y1+beta*y2+4.04*sin(pi/2*t)
    -x(2)-beta*y1+alpha*y2];
```

and the numerical solution is calculated using the following command:

```
[T,Y]=ode45(@twocellsNA,[0:0.001:500],[0.1 0.5],'',2,-1.2);
```

In this case the cell exhibits chaotic behavior. The time series and the attractor on the phase plane $x_1 - x_2$ are reported in Figure 6.11.

Example 6.14

Let us consider the following non-autonomous model:

$$\dot{x}_1 = -x_1 + (1 + \mu + \varepsilon)y_1 - sy_2 + i_1$$
$$\dot{x}_2 = -x_2 + sy_1 + (1 + \mu - \varepsilon)y_2 + i_2$$
(6.33)

Let us assume $\mu = 0.7$, $s = 1$, $\varepsilon = 0$, $i_1 = -0.3$, and $i_2 = 0.3$ and integrate the system dynamics. We define the function `twocellsRD` according to:

```
function dxdt=twocellsRD(t,x,mu,epsi,s,i1,i2)

y1=0.5*(abs(x(1)+1)-abs(x(1)-1));
y2=0.5*(abs(x(2)+1)-abs(x(2)-1));

dxdt=[-x(1)+(1+mu+epsi)*y1-s*y2+i1
    -x(2)+s*y1+(1+mu-epsi)*y2+i2];
```

and then run the command:

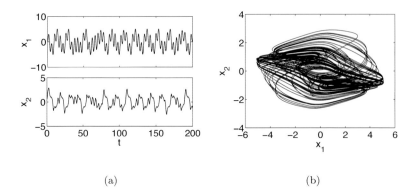

(a)   (b)

FIGURE 6.11
Dynamical behavior of the model (6.32): (a) temporal trend of the two state variables, and (b) chaotic attractor in the phase plane.

```
[T,Y]=ode45(@twocellsRD,[0:0.001:500],[0.1 0.5],'',0.7,0,1,-0.3,0.3);
```

In Figure 6.12 the temporal evolution of the two state variables x_1 and x_2 and the limit cycle described in the phase plane are shown. The oscillations are clearly periodic, but the waveform is far from quasi-sinusoidal.

The dynamical behaviors observed in the three previous examples are sensibly different. While in the second example a chaotic oscillation appears, the first and third examples both display a periodic behavior. However, in the third example a specific feature is observed: oscillations occur with a slow-fast dynamics. This can be clearly seen by looking at Figure 6.13 where the trajectory in the phase plane is reported using markers taken with a constant sampling time. The density of markers is not constant along the cycle, meaning that portions of the cycle are quickly spanned (lower density) while other portions are slowly spanned (higher density). This is a key feature of many reaction-diffusion models and it is the reason why the model in Equations (6.33), characterized by the slow-fast dynamics, is particularly suitable to realize an RD-CNN generating complex phenomena such as pattern formations and autowaves propagation.

6.5.2   Two layer reaction-diffusion CNN

Once Equations (6.33) are identified as a potentially good model to mimic reaction-diffusion equations, we consider them as the dynamical system governing the $M \times N$ cells of a two-dimensional array. The equations of the global system are written as:

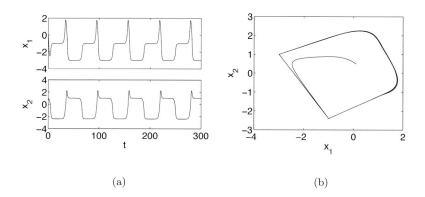

(a)   (b)

FIGURE 6.12
Dynamical behavior of the model (6.33): (a) temporal trend of the two state variables, and (b) trajectory of the system in the phase plane.

$$\dot{x}_1 = -x_1 + (1 + \mu + \varepsilon)y_1 - sy_2 + i_1 +$$
$$D_1(y_{1_{i+1,j}} + y_{1_{i-1,j}} + y_{1_{i,j-1}} + y_{1_{i,j+1}} - 4y_{1_{i,j}})$$
$$\dot{x}_2 = -x_2 + sy_1 + (1 + \mu - \varepsilon)y_2 + i_2 +$$
$$D_2(y_{2_{i+1,j}} + y_{2_{i-1,j}} + y_{2_{i,j-1}} + y_{2_{i,j+1}} - 4y_{2_{i,j}})$$

$$(6.34)$$

with $1 \leq i \leq M$ and $1 \leq j \leq N$. The system represents an RD-CNN, and, in particular, a two-layer CNN. The interactions with the neighboring cells are obtained by means of two diffusive terms acting along the two layers with diffusion coefficients D_1 and D_2. The model is reformulated in the standard two-layer CNN form as:

$$\dot{\mathbf{x}}_{ij} = -\mathbf{x}_{ij} + \mathbf{A} * \mathbf{y}_{ij} + \mathbf{I} \tag{6.35}$$

where $\mathbf{x}_{ij} = \begin{bmatrix} x_{1_{i,j}} & x_{2_{i,j}} \end{bmatrix}^T$, $\mathbf{y}_{ij} = \begin{bmatrix} y_{1_{ij}} & y_{2_{ij}} \end{bmatrix}^T$, and $\mathbf{I} = \begin{bmatrix} I_1 & I_2 \end{bmatrix}^T$.

For this system the feedback cloning template is given by

$$\mathbf{A} = \begin{bmatrix} A_{11} & A_{12} \\ A_{21} & A_{22} \end{bmatrix}$$

with:

$$A_{11} = \begin{bmatrix} 0 & D_1 & 0 \\ D_1 & -4D_1 + \mu + 1 & D_1 \\ 0 & D_1 & 0 \end{bmatrix}$$

$$A_{12} = \begin{bmatrix} 0 & 0 & 0 \\ 0 & -s & 0 \\ 0 & 0 & 0 \end{bmatrix}$$

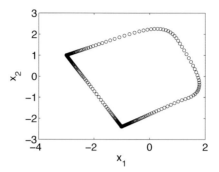

FIGURE 6.13
Trajectory of system (6.33): slow-fast dynamics is emphasized by markers taken with constant sampling time.

$$A_{21} = -A_{12}$$

$$A_{22} = \begin{bmatrix} 0 & D_2 & 0 \\ D_2 & -4D_2 + \mu + 1 & D_2 \\ 0 & D_2 & 0 \end{bmatrix}$$

Zero-flux boundary conditions are assumed.

This reaction-diffusion CNN can be considered a reference for modeling, by means of circuits, different processes, and phenomena. In fact, it can emulate both autowaves and Turing patterns.

The simulation of an RD-CNN can be easily realized implementing a specific MATLAB® routine slightly modifying the one introduced for the image processing. The main difference relies on the fact that the RD-CNN is a two-layer structure. In the following examples, the generation of spiral waves, autowaves, Turing patterns, and noisy autowaves will be discussed using the following MATLAB® function:

```
function [X1,Y1,X2,Y2,X1d,Y1d,X2d,Y2d]=cnn2D(X1i,X2i,A11,A22,I1,I2,miu,
    epsi,s,dt,step)
[N1,M1]=size(X1i);
X1i=normalizeCNN(X1i);
X2i=normalizeCNN(X2i);
N=N1+2;
M=M1+2;
X1=zeros(N,M,step);
X2=zeros(N,M,step);
Y1=zeros(N,M,step);
Y2=zeros(N,M,step);
X1(2:end-1,2:end-1,1)=X1i;
X2(2:end-1,2:end-1,1)=X2i;
Y1(:,:,1)=0.5*(abs(X1(:,:,1)+1)-abs(X1(:,:,1)-1));
Y2(:,:,1)=0.5*(abs(X2(:,:,1)+1)-abs(X2(:,:,1)-1));
for t=2:step
    for i=2:(N1+1)
        for j=2:(M1+1)
            coup1A=0;
```

```
              coup2A=0;
              for k=[-1 0 1]
                  for l=[-1 0 1]
                      coup1A=coup1A+A11(k+2,l+2)*Y1(i+k,j+1,t-1);
                      coup2A=coup2A+A22(k+2,l+2)*Y2(i+k,j+1,t-1);
                  end
              end
              X1(i,j,t)=X1(i,j,t-1)+dt*(-X1(i,j,t-1)+(1+miu+epsi)*Y1(i,j,t-1)
                  -s*Y2(i,j,t-1)+coup1A+I1);
              Y1(i,j,t)=0.5*(abs(X1(i,j,t)+1)-abs(X1(i,j,t)-1));
              X2(i,j,t)=X2(i,j,t-1)+dt*(-X2(i,j,t-1)+s*Y1(i,j,t-1)
                  +(1+miu-epsi)*Y2(i,j,t-1)+coup2A+I2);
              Y2(i,j,t)=0.5*(abs(X2(i,j,t)+1)-abs(X2(i,j,t)-1));
          end
      end
      [X1t,Y1t]=Boundary(squeeze(X1(:,:,t)),squeeze(Y1(:,:,t)),2);
      [X2t,Y2t]=Boundary(squeeze(X2(:,:,t)),squeeze(Y2(:,:,t)),2);
      X1(:,:,t)=X1t;
      X2(:,:,t)=X2t;
      Y1(:,:,t)=Y1t;
      Y2(:,:,t)=Y2t;
end
X1d=denormalizeCNN(X1(:,:,end));
Y1d=denormalizeCNN(Y1(:,:,end));
X2d=denormalizeCNN(X2(:,:,end));
Y2d=denormalizeCNN(Y2(:,:,end));
X1d=X1d(2:end-1,2:end-1);
Y1d=Y1d(2:end-1,2:end-1);
X2d=X2d(2:end-1,2:end-1);
Y2d=Y2d(2:end-1,2:end-1);
end
```

where the functions `normalizeCNN`, `denormalizeCNN`, and `Boundary` are those reported above.

Example 6.15

Generation of a spiral wave.

To generate a spiral wave in a two-layer CNN, let us consider as initial states of the two layers the two gray-scale images reported in Figure 6.14. They represent the seed of a single wavefront. The RD-CNN (6.34) is simulated with $\mu = 0.7$, $s = 1$, $\varepsilon = 0$, $i_1 = -0.3$, $i_2 = 0.3$, and $D_1 = D_2 = 0.1$. The following command is used:

```
[X1,Y1,X2,Y2,X1d,Y1d,X2d,Y2d]=cnn2D(X1i,X2i,
    [0 0.1 0;0.1 -4*0.1 0.1;0 0.1 0],
    [0 0.1 0;0.1 -4*0.1 0.1;0 0.1 0],
    -0.3,0.3,0.7,0,1,0.1,700);
```

The result is a spatiotemporal phenomenon of propagation of a spiral wave in the structure. A snapshot at time $T = 700$ of the output in the two layers is shown in Figure 6.15.

Example 6.16

Generation of Turing patterns.

The generation of complex patterns in a two-layer CNN can be obtained by suitably choosing the parameters of the model. With reference to the RD-CNN in Equations (6.34) we can set $\mu = -0.1$, $s = 2$, $\varepsilon = 2$, $i_1 = i_2 = 0$, $D_1 = 0.01$, $D_2 = 1$. We consider a 40×40 CNN. The initial state for the first layer is the image reported in Figure 6.16(a), while the same image but with reversed sign is used to fix the initial conditions for the second layer. The following commands are used to perform the simulation:

```
[X1,Y1,X2,Y2,X1d,Y1d,X2d,Y2d]=cnn2D(X1i,X2i,
    [0 0.01 0;0.01 -4*0.01 0.01;0 0.01 0],
```

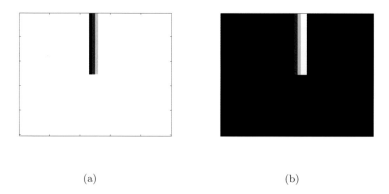

(a)   (b)

FIGURE 6.14
Initial states of the two layers of the RD-CNN for the generation of a spiral
wave: (a) $x_{1_{i,j}}(0)$, (b) $x_{2_{i,j}}(0)$.

```
[0 1 0;1 -4*1 1;0 1 0],
0,0,-0.1,2,2,0.1,1000);
```

The result is a steady-state pattern (a Turing pattern) which is shown in Figure 6.16(b).

Example 6.17 ⎯⎯⎯⎯⎯⎯⎯⎯⎯⎯⎯⎯⎯⎯⎯⎯⎯⎯⎯⎯⎯⎯⎯⎯⎯⎯⎯⎯⎯⎯⎯⎯⎯⎯

Generation of a spiral wave in a noisy active medium.
The process which drives the generation of an autowave is particularly robust to
noise in the active medium. To verify this, we performed the following experiment.
Consider as initial states of the two layers of the RD-CNN the two pictures shown
in Figure 6.17(a)-6.17(b). The two images have been generated adding a Gaussian
noise to the images of Figure 6.14. The RD-CNN (6.34) is simulated with $\mu = 0.7$,
$s = 1$, $\varepsilon = 0$, $i_1 = -0.3$, $i_2 = 0.3$, and $D_1 = D_2 = 0.1$. The following commands
are used:

```
[X1,Y1,X2,Y2,X1d,Y1d,X2d,Y2d]=cnn2D(X1i,X2i,
    [0 0.1 0;0.1 -4*0.1 0.1;0 0.1 0],
    [0 0.1 0;0.1 -4*0.1 0.1;0 0.1 0],
    -0.3,0.3,0.7,0,1,0.1,700);
```

As can be observed from the output of the two layers at time $T = 300s$, shown in
Figure 6.17(c)-6.17(d), the spiral wave is subjected to spurious wavefronts generated by the noise. However, after some time, these spurious wavefronts are finally
annihilated by the main spiral wavefront, as reported in Figure 6.17(e)-6.17(f).

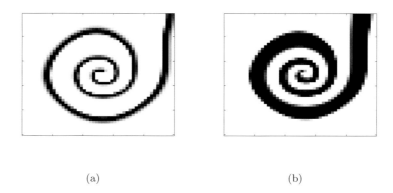

(a) (b)

FIGURE 6.15
A snapshot of the spiral wave propagating in the two layers of the RD-CNN at time $T = 700$: (a) $y_{1_{i,j}}(T)$, (b) $y_{2_{i,j}}(T)$.

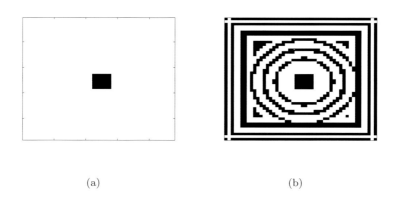

(a) (b)

FIGURE 6.16
Generation of Turing patterns: (a) initial states $x_{1_{i,j}}(0)$, (b) Turing pattern obtained at time $T = 1000$ on layer $y_{1_{i,j}}(T)$.

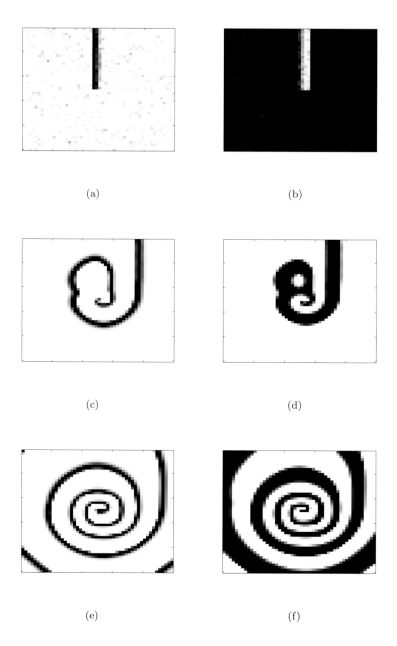

(a)   (b)

(c)   (d)

(e)   (f)

FIGURE 6.17
Spiral wave propagating in a noisy active medium emulated by an RD-CNN:
(a) $x_{1_{i,j}}(0)$, (b) $x_{2_{i,j}}(0)$; (c) layer $y_{1_{i,j}}(T)$ at $T = 300s$; (d) layer $y_{2_{i,j}}(T)$ at
$T = 300s$; (e) layer $y_{1_{i,j}}(T)$ at $T = 700s$; (f) layer $y_{2_{i,j}}(T)$ at $T = 700s$.
Defects and spurious wavefronts are initially present and then are annihilated
by the main spiral wavefront.

6.5.3 Chua's circuit reaction-diffusion CNN

The Chua's circuit realized with an SC-CNN is a three-layer CNN where each layer has one cell. Let us consider the scheme introduced in Figure 6.18. It represents a very simple reaction-diffusion CNN circuit obtained coupling two three-layer CNNs (each emulating the Chua's circuit dynamics) with a diffusive resistor R_c. The diffusive coefficient D is given by $\frac{\alpha R_6}{R_c}$, while the reactive term is the Chua's circuit dynamics. The equations of the two diffusively coupled Chua's circuits are:

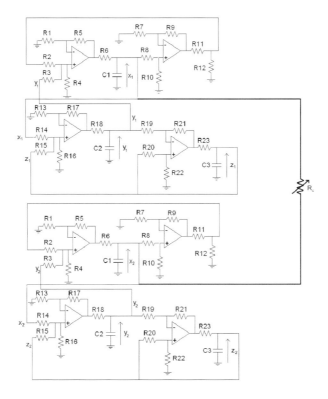

FIGURE 6.18
RD-CNN with two cells, each implementing the Chua's circuit dynamics.

$$\dot{x}_1 = \alpha[y_1 - h(x_1)] + D(x_2 - x_1)$$
$$\dot{y}_1 = x_1 - y_1 + z_1$$
$$\dot{z}_1 = -\beta y_1$$
$$\dot{x}_2 = \alpha[y_2 - h(x_2)] + D(x_1 - x_2) \qquad (6.36)$$
$$\dot{y}_2 = x_2 - y_2 + z_2$$
$$\dot{z}_2 = -\beta y_2$$

We remark that the two diffusively coupled three-layer CNNs shown in Figure 6.18 can be easily experimentally realized. Indeed, in Chapter 8, experimental results referred to this system will be discussed.

The previous scheme can be generalized in two ways. First, the diffusion may take place also among other variables and, therefore, more diffusive elements may be considered. Moreover, arrays of circuits can be coupled by using diffusive elements. Therefore 1D, 2D, 3D arrays can be taken into account.

Example 6.18 _____

Let us write the equations of a 3D RD-CNN of Chua's circuits with two diffusive elements between variables x and y. The equations are the following:

$$\dot{x}_{i,j,k} = \alpha[y_{i,j,k} - h(x_{i,j,k})] +$$
$$\quad D_x[x_{i+1,j,k} + x_{i-1,j,k} + x_{i,j+1,k} + x_{i,j-1,k} + x_{i,j,k+1} + x_{i,j,k-1} - 6x_{i,j,k}]$$
$$\dot{y}_{i,j,k} = x_{i,j,k} - y_{i,j,k} + z_{i,j,k} +$$
$$\quad D_y[y_{i+1,j,k} + y_{i-1,j,k} + y_{i,j+1,k} + y_{i,j-1,k} + y_{i,j,k+1} + y_{i,j,k-1} - 6y_{i,j,k}]$$
$$\dot{z}_{i,j,k} = -\beta y_{i,j,k}$$

$$(6.37)$$

The diffusion occurs at each layer in the three spatial dimensions, as schematized in Figure 6.19. By using the previous model and the corresponding circuits it has been proved that Turing patterns, autowaves, and spiral waves can be obtained.

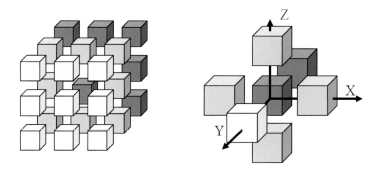

FIGURE 6.19

Schematic representation of a 3D RD-CNN in which diffusion occurs along the three spatial dimensions.

6.5.4 Reaction-diffusion CNN as a network of cells

Moving away from the Chua's circuit dynamics, we may refer to a plethora of reaction-diffusion equations that can be found in the literature. These equations are used to model various physical processes such as chemical reactions or action potentials propagating in a network of neurons. A non-exhaustive set of case studies of reaction-diffusion systems and a few of the behaviors that can be observed in such systems are illustrated with the following examples. Following the discussion presented in this section, all these examples can be modeled by using multilayer RD-CNNs.

Example 6.19 _____

FitzHugh–Nagumo equations [25].
This model describes the propagation of action potentials along axons in a large population of neurons. The equations are as follows:

$$\begin{aligned} \frac{\partial x}{\partial t} &= -\left(\frac{x^3}{3} - x\right) - y + D_1 \nabla^2 x \\ \frac{\partial y}{\partial t} &= -\varepsilon(x - by) + D_2 \nabla^2 y \end{aligned} \tag{6.38}$$

where x represents the membrane potential and y is a recovery variable. A 100×100 RD-CNN in which each cell is described by Equations (6.38) with $b = 1.5$, $\varepsilon = -0.1$, $D_1 = 0.3$, and $D_2 = 1.4$ has been simulated and the spot-like pattern shown in Figure 6.20 emerges in the layer of the recovery variable y after an integration time $T = 100s$.

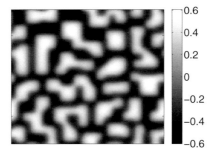

FIGURE 6.20
RD-CNN of 100×100 FitzHugh–Nagumo cells: pattern on the layer y at $T = 100s$.

Example 6.20 _____

Brusselator equations [26].
This model has been proposed by Prigogine and Lefever to describe autocatalytic reactions. The Brusselator equations read:

$$\begin{aligned} \frac{\partial x}{\partial t} &= a - (b + 1)x + x^2 y + D_1 \nabla^2 x \\ \frac{\partial y}{\partial t} &= bx - xy + D_2 \nabla^2 y \end{aligned} \tag{6.39}$$

where x and y represent the concentration of the two autocatalytic species involved in the reaction. A 100×100 RD-CNN in which each cell is described by Equations (6.39) with $a = -0.5$, $b = 1.5$, $D_1 = 1$, and $D_2 = 0$ has been simulated and the ordered spot pattern shown in Figure 6.21 can be found in both layers after an integration time of $T = 100s$.

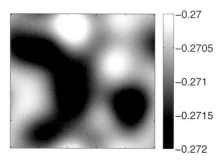

FIGURE 6.21
RD-CNN of 100×100 Brusselator cells: pattern on the layer x at $T = 100s$.

Example 6.21 _____

Gierer–Meinhardt model [27].
The model describes the activator-inhibitor interplay that appears in many important types of pattern formation and morphogenesis. The model is governed by the following set of equations:

$$\begin{aligned} \frac{\partial x}{\partial t} &= a + \frac{x^2}{y} - bx + D_1\nabla^2 x \\ \frac{\partial y}{\partial t} &= x^2 - y + D_2\nabla^2 y \end{aligned} \tag{6.40}$$

where x is the concentration of the short-range autocatalytic species (the activator) and y is the concentration of the long-range antagonist (the inhibitor). A 100×100 RD-CNN in which each cell is described by Equations (6.40) with $a = -0.5$, $b = 0.5$, $D_1 = 1$, and $D_2 = 3$ has been simulated and the pattern shown in Figure 6.22 is found in the activator layer at time $T = 100s$.

Example 6.22 _____

Oregonator equations [58]
Similarly to the Brusselator equations, this model represents the dynamics of autocatalytic reactions. In particular, it is able to model the so-called Belusov–Zhabotinsky reaction. The three equations used to describe the model are:

$$\begin{aligned} \frac{\partial x}{\partial t} &= x + y - \alpha x^2 - xy + D_1\nabla^2 x \\ \frac{\partial y}{\partial t} &= -y + \beta z - xy + D_2\nabla^2 y \\ \frac{\partial z}{\partial t} &= x - z + D_3\nabla^2 z \end{aligned} \tag{6.41}$$

where x, y, and z represent the concentrations of the three reactants. A 100×100 RD-CNN in which each cell is described by Equations (6.41) with $\alpha = 3$, $\beta = 3$, $D_1 = 1$, and $D_2 = D_3 = 0$ has been simulated and the pattern shown in Figure 6.23 is found in the first layer at time $T = 100s$.

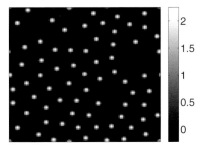

FIGURE 6.22
RD-CNN of 100×100 Gierer–Meinhardt cells: pattern on the layer x at $T = 100s$.

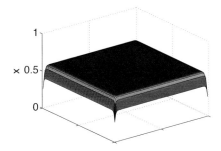

FIGURE 6.23
RD-CNN of 100×100 Oregonator cells: pattern on the layer x at $T = 100s$.

The idea arising from the previous examples is that a reaction-diffusion equation can be conceived as an electronic circuit, thus exploiting circuit theory and, in particular, the CNN concept. In fact, considering each cell as the reactive term and the local propagation phenomenon, that is the peculiarity of the CNN architecture, as the diffusive term, an evident dichotomy between the reaction-diffusion equations and the CNN paradigm appears. We conclude that the cell is the primitive network of complex systems and that these systems can be formulated within the CNN paradigm.

6.5.5 Diffusive networks of multilayer CNN

We can generalize the CNN scheme to the case where we have networks of n-dimensional dynamical systems, each of which can be assumed as a cluster of more than one subsystem, as schematically represented in Figure 6.24. Each subsystem, in fact, can be arranged as a multilayer CNN. It is evident that the dynamical behavior of a system of networks emerges by the cooperative behavior of a large number of units that we call cells. If we imagine that each cell can be realized by means of an electronic elementary nonlinear circuit and clusters of cells are linked together, a complex dynamical system can be viewed as a network of a large number of cells.

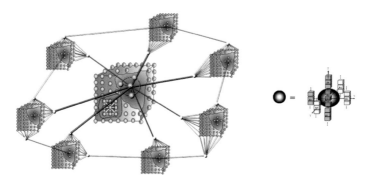

FIGURE 6.24
Schematic representation of a network of clusters of subsystems. Each subsystem can be represented as a multilayer CNN.

A particular case will be experimentally investigated in Chapter 8. Up to now identical cells and local diffusive elements have been considered. In the network experimentally realized, each unit is a multilayer CNN implementing the Chua's circuit dynamics but the possibility to have parametric differences is also considered, so that the dynamical behavior shown by each cell may be different. Furthermore, the diffusive effect is not limited to be local, but long-range connections or complex topologies are also taken into account. This is a new scheme of network that includes diversity both in the unit dynamics and in the topological scheme and constitutes a further generalization of the classical CNN paradigm.

6.6 Summary

In this chapter the main concepts of CNNs have been given. The various issues have been described with the aim of focusing the attention of the reader on the possibility offered by CNNs to conceive analog circuits emulating complex dynamics. The concept of cell has been widely emphasized and ideas that will be useful for discussing the applications of the second part of the book have been developed.

Even if high level technology is at the basis of VLSI CNN devices, the paradigm of CNN makes possible their realization for practical applications with low-cost components allowing the readers to realize CNNs in educational electronic laboratories.

The concept has been widely generalized and the various topologies of CNNs have been illustrated. The role of CNNs as analog electronic coprocessors has been also remarked upon.

6.7 Exercises

1. Consider the standard two-dimensional CNN model with $R_x C = 1$. Write the equations of a 3×3 CNN considering parametric the cloning templates.

2. For the CNN model in the previous exercise, consider the pattern reported in Figure 6.25(a): find the values of the entries of the cloning templates so that the output of the CNN is that reported in Figure 6.25(b).

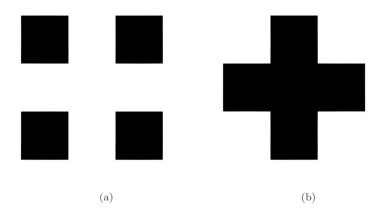

(a)   (b)

FIGURE 6.25
Exercise 2: (a) input image, and (b) desired output image.

3. Consider the following set of cloning templates:

$$A = \begin{bmatrix} 0 & 0 & 0 \\ 0 & 3 & 0 \\ 0 & 0 & 0 \end{bmatrix} ; \quad B = \begin{bmatrix} 0.25 & 0 & 0 \\ 0 & 0 & 0 \\ 0 & 0 & 0.25 \end{bmatrix} ; \quad I = -1 \qquad (6.42)$$

Verify the output of the CNN when considering Figure 6.26 as input and initial condition, with null fixed boundary conditions.

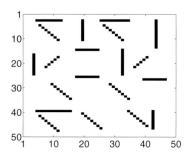

FIGURE 6.26
Input and initial condition image for Exercise 3.

4. Consider a 1D CNN where $A = \begin{bmatrix} \bar{a} & \bar{b} & a \end{bmatrix}$. Find through simulations the values of the parameters $\bar{a}$, $\bar{b}$, and a so that the CNN is stable.

5. Consider the non-autonomous chaotic CNN model given in Equations (6.32) and calculate the Poincaré section of its dynamics.

6. Use a CNN to extract the edges of the Poincaré section retrieved in the previous exercise.

7. Consider the following reaction-diffusion equations:

$$\frac{\partial A}{\partial t} = c_1 A + c_2 I + D_A \frac{\partial^2 A}{\partial x^2}$$
$$\frac{\partial I}{\partial t} = c_3 A + c_4 I + D_I \frac{\partial^2 I}{\partial x^2} \qquad (6.43)$$

Derive the equivalent CNN model.

8. Consider the activator-inhibitor system represented by the following equations:

$$\frac{\partial a}{\partial t} = s\left(\frac{a^2}{b} + ba\right) - r_a a + D_a \frac{\partial^2 a}{\partial x^2}$$
$$\frac{\partial b}{\partial t} = sa^2 - r_b b + b + D_b \frac{\partial^2 b}{\partial x^2} \qquad (6.44)$$

Derive the equivalent CNN model.

9. Consider the following model of autocatalysis:

$$\begin{array}{l}
\frac{\partial a}{\partial t} = \frac{s}{s_a + c^2} - r_a a + ba + D_a \frac{\partial^2 a}{dx^2} \\
\frac{\partial b}{\partial t} = r_b a - r_b b + D_b \frac{\partial^2 b}{dx^2} \\
\frac{\partial c}{\partial t} = \frac{s}{s_c + \frac{a^2}{b^2}} - r_c c + D_c \frac{\partial^2 c}{dx^2}
\end{array} \tag{6.45}$$

Derive the equivalent CNN model.

Further reading

For additional information on the topics of the chapter, the following references may be consulted: [20], [41], [56].

7

Synchronization and chaos control

CONTENTS

The characteristics of chaos and the strategies to implement nonlinear circuits exhibiting chaotic behavior have been dealt with in the previous chapter. The aim of this chapter is to control chaotic dynamics, despite its long-term unpredictable behavior and its sensitivity to perturbation on parameters and initial conditions. The possibility to control chaos is challenging and of particular interest from the circuit point of view. We will introduce the concept of chaos synchronization, that is the condition under which two (or many) chaotic systems may follow the same trajectory even starting from different initial conditions, and the concept of chaos control, that is the intentional induction of a chaotic behavior (or the suppression of it) by means of an external action.

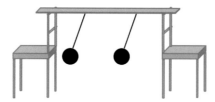

FIGURE 7.1
Schematic illustration of the Huygens' experiment on synchronization.

7.1   Introduction

According to the Britannica English dictionary, the meaning of the term "to synchronize" is "to occur at the same time", "to make contemporaneously"; the term "synchronism" indicates "the state of being synchronous" and "synchronous" means "occurring at the same time", "to be coincident", "happening at the same time". From a physical point of view this means to have the same period or rate of oscillation.

The first studies on synchronization are historically attributed to the Dutch scientist Christian Huygens. He observed that two pendulum clocks anchored to the same beam supported by two chairs (schematically illustrated in Figure 7.1), after a while (about half an hour in his observations), swing at synchronous in anti-phase. He also noticed that different initial conditions did not affect the synchronous motion of the two pendulums. Huygens named this effect "sympathie des horloges". The phenomenon is now known as synchronization and represents the adjustment of the rhythm of the oscillations of two or more systems due to the weak interactions between them.

Studying synchronization in dynamical systems provides important insights for the understanding and modeling of a phenomenon that is ubiquitous in natural and man-made systems [71, 5]. There are many natural systems that display synchronization in an often fascinating way: the unison song of crickets, the synchronous rhythmic flash of fireflies observed in Bornco forests, the spontaneous synchronization of clapping in a human platea are a few limited examples. Concerning man-made systems, there are in fact devices where the synchronous behavior leads to an enhancement of the performance. An example is an array of Josephson junctions where higher output power values are obtained when the junctions oscillate synchronously. Hence, the study of synchronization in complex systems has a twofold importance: from the theoretical point of view it allows us to better understand natural phenomena, and from a technological point of view it is useful to design high-performance devices and systems.

Synchronization is the process through which two or more dynamical systems (that can be identical or not) coupled through some form of interaction or driven by a common forcing adjust a particular property of their motion (this could be, for instance, the frequency or the phase of the oscillation or the complete trajectory). From the Huygens's observations two factors having a key role in the phenomenon are in evidence: the weak interaction between the two clocks and their different initial conditions.

Although the occurrence of synchronization has been originally studied in systems of coupled periodic oscillators, in this chapter our attention is devoted to the main concepts regarding synchronization of chaotic systems, including as a special case that of periodic oscillators. The chapter also illustrates some key ideas on chaos control, a topic that is strictly related to that of synchronization. Unlike synchronization that is also important for periodic systems, chaos control is clearly a topic specific only to circuits and systems that can exhibit chaotic dynamics. However, the concept of weak interactions is relevant also for this topic. In fact, it can be considered as a requirement for the control action to be designed. As previously outlined, chaotic behavior is associated to strange attractors that contain an infinite number of unstable limit cycles. One of the questions to which chaos control tries to provide an answer is whether it is possible to stabilize, through a weak control action, one of these orbits making periodic a system otherwise chaotic. On the other hand, in some cases it is useful to induce a chaotic behavior in systems that without control are not chaotic. This latter case is the problem of anti-chaos control.

7.2 Principles of synchronization of nonlinear dynamical systems

Consider two nonlinear circuits (chaotic or not), as in Figure 7.2, that interact with each other. The arrows indicate the direction of coupling, so that in the case shown in Figure 7.2 the interaction is bidirectional as in the Huygens experiment. The coupling may also be unidirectional, with one circuit (the driver or the master) driving the other (the response system or the slave).

Let $\mathbf{x}$ indicate the state vector of the first circuit and $\mathbf{y}$ that of the second circuit. There exist different forms of synchronization that can occur between two nonlinear systems. The first one represents the strongest condition and is referred to as complete synchronization. It occurs when the state variables of the two systems asymptotically follow the same trajectory, that is the two systems are said to be completely synchronized if

$$\lim_{t \to \infty} \|\mathbf{x}(t) - \mathbf{y}(t)\| = 0 \qquad (7.1)$$

FIGURE 7.2
Two bidirectionally coupled nonlinear circuits.

for any initial conditions of the two circuits in the basin of attraction of the synchronization manifold, where $\| \ \|$ denotes the Euclidean norm of a vector.

Our attention is devoted to the case in which both circuits exhibit chaotic behavior. In this case, if condition (7.1) is satisfied, despite the sensitivity to initial conditions, the two signals $\mathbf{x}(t)$ and $\mathbf{y}(t)$ after some time will have the same trend. The sensitivity to initial conditions of chaotic circuits makes this problem non-trivial as, even if the systems are identical, without any interaction they will follow different trajectories. The fascinating property of synchronization is that a weak coupling makes possible a common evolution of the two systems.

The phenomenon of synchronization manifests in many other forms beyond the complete asymptotic identity of the state variables of two systems. For instance, synchronization can regard only a property of the trajectory, for example the phase. In this case we say that there is phase locking or phase synchronization between two systems. More precisely, this occurs when, said $\varphi_1(t)$ and $\varphi_2(t)$ the phases of the two systems, it holds that:

$$|\varphi_1(t) - \varphi_2(t)| < \text{const} \quad \forall t > t_0 \tag{7.2}$$

that is the phase difference remains bounded for any time. Despite the fact that the definition of the phase is rather intuitive for a quasi-linear oscillator (as its oscillations resemble a sinusoidal waveform), for a chaotic system the problem is non-trivial and there exist various definitions. If for a chaotic circuit a certain reference point around which the trajectory of the chaotic system rotates can be defined, then the phase can be defined as $\varphi(t) = \arctan \frac{x_2(t)}{x_1(t)}$, where $x_1(t)$ and $x_2(t)$ are the coordinates of a reference system such that the rotation reference point is the origin and the attractor projected on the plane $x_1 - x_2$ evidences the rotations around the origin. More in general, to define the phase of a chaotic system one has to rely on proper mathematical tools such as the Hilbert transform. In some cases an appropriate Poincaré section may also be used for the purpose.

The condition (7.2) implies that the frequencies defined as $\omega_i = \dot{\varphi}_i$ of the systems are also locked. The condition (7.2) may be also generalized to account for more complex patterns of synchronization:

$$|n\varphi_1(t) - m\varphi_2(t)| < \text{const} \quad \forall t > t_0 \tag{7.3}$$

with n and m integer numbers. In this case there is $n : m$ phase synchronization, that is, the frequencies are locked such that $n\omega_1 = m\omega_2$. This form of synchronization can be easily detected on the oscilloscope using the xy mode, that is, reporting one signal as a function of the other. One obtains a Lissajous figure revealing this type of locking. An 8-shaped curve, for instance, indicates a $1 : 2$ synchronization since to each cycle of one oscillator correspond two complete cycles of the other.

Lag synchronization is instead established when two systems have nearly identical oscillations, but one lags in time with respect to the other. Finally, generalized synchronization is the most general, and difficult to detect, form of synchronization. Suppose that the flow of the interaction is from system 1 with state variables $\mathbf{x}$ to system 2 with state variables $\mathbf{y}$. Generalized synchronization is connected to the existence of a mapping $\psi : \mathbf{x} \to \mathbf{y}$, that can be asymptotically established transforming the trajectory of the driver system into that of the response system, $\mathbf{y}(t) = \psi(\mathbf{x}(t))$. The presence of such mapping clearly indicates that system 2 is no more independent from system 1, or equivalently that the evolution of system 2 can be predicted from that of system 1. Generalized synchronization can be very difficult to detect and appropriate methods and indicators, some of them based on statistical analysis of the signals, are available in the literature.

An aspect that is very interesting is the transition from one form of synchronization to another. Very often as the coupling strength is increased from zero, first some forms of weak synchronization appear (generalized or phase synchronization for instance) that then may transform into a stronger one (complete synchronization for instance). We also mention that, when more than two units are coupled into a network configuration, a plethora of disparate patterns of synchronization may occur. Chimera states, relay and remote synchronization are only a few examples. For a deeper discussion the reader is referred to the literature on the topic [1, 4, 29, 33, 38, 64, 66].

For the remaining part of the chapter we will restrict our discussion to complete synchronization of chaotic systems. First unidirectional coupling and, then, bidirectional coupling will be dealt with.

7.3 Schemes for unidirectional synchronization

In this section several methods to achieve synchronization in unidirectionally coupled systems are presented. The circuit sending the signal for synchronization is referred to as the *master* system, while that receiving the signal is the *slave* system. Synchronization obtained with these schemes is referred to as master-slave synchronization or as drive-response synchronization.

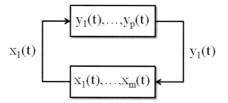

FIGURE 7.3
Decomposition of a system into two blocks with scalar interactions between
them.

7.3.1 Master-slave synchronization by system decomposition

Suppose that the master and slave systems may be decomposed as follows:

$$
\begin{aligned}
\dot{x}_1 &= f_1(x_1, \ldots, x_m, y_1) \\
\dot{x}_2 &= f_2(x_1, \ldots, x_m, y_1) \\
&\cdots \\
\dot{x}_m &= f_m(x_1, \ldots, x_m, y_1) \\
\dot{y}_1 &= g_1(x_1, y_1, \ldots, y_p) \\
\dot{y}_2 &= g_2(x_1, y_1, \ldots, y_p) \\
&\cdots \\
\dot{y}_p &= g_p(x_1, y_1, \ldots, y_p)
\end{aligned}
\tag{7.4}
$$

that is, the set of n state variables of the master or of the slave is decomposed into two subsets, one with m variables $x_1, \ldots, x_m$ and one with p variables $y_1, \ldots, y_p$, where the dynamics of $x_1, \ldots, x_m$ depend on y_1 (but not on $y_2, \ldots, y_p$); and, vice versa, the dynamics of $y_1, \ldots, y_p$ depend on x_1 (but not on $x_2, \ldots, x_m$). This decomposition results in the scheme of Figure 7.3.

Based on this decomposition, the synchronization scheme of Figure 7.4 is designed. The slave is forced by the master through the signal $y_1(t)$. Under the condition that the conditional Lyapunov exponents are negative (the conditional Lyapunov exponents are the Lyapunov exponents of the slave system subjected to the driving of the master, considered an external input signal) this scheme leads to synchronization. We note however that this scheme uses a form of strong coupling.

Example 7.1 _____

As an example of synchronization based on system decomposition [68], we consider the Lorenz system:

$$
\begin{aligned}
\dot{X} &= \sigma(Y - X) \\
\dot{Y} &= \rho X - Y - XZ \\
\dot{Z} &= XY - bZ
\end{aligned}
\tag{7.5}
$$

We note that the system (7.5) can be decomposed as in Equation (7.4) with $x_1 = X$, $y_1 = Y$, and $y_2 = Z$:

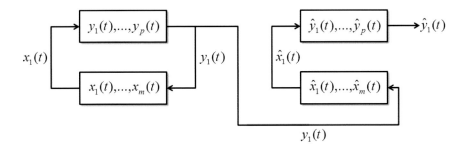

FIGURE 7.4
Master-slave synchronization based on system decomposition.

$$\dot{x}_1 = \sigma(y_1 - x_1)$$
$$\dot{y}_1 = \rho x_1 - y_1 - x_1 y_2 \qquad (7.6)$$
$$\dot{y}_2 = x_1 y_1 - b y_2$$

Equation (7.6) represents the master. We design the synchronization scheme by sending the variable x_1. This variable will be the input of the slave system, that is $u = x_1$, whose equations are written as:

$$\dot{\hat{x}}_1 = \sigma(\hat{y}_1 - \hat{x}_1)$$
$$\dot{\hat{y}}_1 = \rho x_1 - \hat{y}_1 - x_1 \hat{y}_2 \qquad (7.7)$$
$$\dot{\hat{y}}_2 = x_1 \hat{y}_1 - b \hat{y}_2$$

It is now proven that the slave system synchronizes with the master. To this aim, we define the errors $e_1(t) = x_1(t) - \hat{x}_1(t)$, $e_2(t) = y_1(t) - \hat{y}_1(t)$, and $e_3(t) = y_2(t) - \hat{y}_2(t)$, and compute the error dynamics:

$$\dot{e}_1 = \sigma(e_2 - e_1)$$
$$\dot{e}_2 = -e_2 - x_1 e_2 \qquad (7.8)$$
$$\dot{e}_3 = x_1 e_1 - b e_2$$

The second Lyapunov theorem is now applied to prove that the equilibrium point $(e_1, e_2, e_3) = (0, 0, 0)$ of system (7.8) is stable. Once this is proved, synchronization immediately follows, as $\lim_{t \to \infty} e_h(t) = 0$ $(h = 1, 2, 3)$ implies that $\hat{x}_1 \to x_1$, $\hat{y}_1 \to y_1$, and $\hat{y}_2 \to y_2$.

To apply the second Lyapunov theorem, consider the Lyapunov function

$$V(e_1, e_2, e_3) = \frac{1}{2}\left(\frac{1}{\sigma}e_1^2 + e_2^2 + e_3^2\right) \qquad (7.9)$$

$V(e_1, e_2, e_3)$ is positive definite in $(e_1, e_2, e_3) = (0, 0, 0)$. Consider now $\dot{V} = \frac{dV}{dt}$:

$$\dot{V} = \frac{1}{\sigma}e_1\dot{e}_1 + e_2\dot{e}_2 + e_3\dot{e}_3 \qquad (7.10)$$

Taking into account the error dynamics (7.8), we obtain

$$\dot{V} = -(e_1 - \frac{1}{2}e_2)^2 - \frac{3}{4}e_2^2 - b e_3^2 \qquad (7.11)$$

that is negative definite, thus proving the stability of the origin for the error dynamics and the suitability of the adopted system decomposition scheme for synchronization.

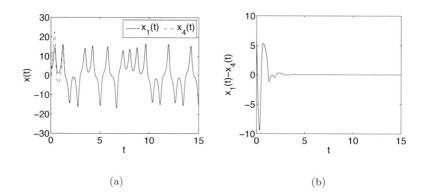

(a)   (b)

FIGURE 7.5
Master-slave synchronization via system decomposition of two Lorenz systems:
(a) trends of $x_1(t)$ and $x_4(t)$ (these MATLAB® variables correspond to $x_1(t)$
and $\hat{x}_1(t)$ in Equations (7.6) and (7.7)); (b) difference $x_1(t) - x_4(t)$.

To simulate the configuration, we write Equations (7.6) and (7.7) in the file
mssysdeclorenzeqs.m:

```
function dxdt = mssysdeclorenzeqs(t,x)

rho = 28;
sigma = 10;
beta = 8/3;

dxdt = [sigma*(x(2) - x(1));
    rho*x(1)- x(2)-x(1)*x(3);
    x(1)*x(2) - beta*x(3)
    sigma*(x(5) - x(4));
    rho*x(1)- x(5)-x(1)*x(6);
    x(1)*x(5) - beta*x(6)];
```

and integrate them and visualize the solution with the commands:

```
[t,x]=ode45(@mssysdeclorenzeqs,[0 15],[5 0.7 20 -1 2 1]);
figure, plot(t,x(:,1:3:end)), xlabel('t'), ylabel('x(t)')
figure, plot(t,x(:,1)-x(:,4)), xlabel('t'), ylabel('x_1(t)-x_4(t)')
```

Figure 7.5 shows the results, illustrating how synchronization is attained after a
short transient. The two variables $x_1(t)$ and $\hat{x}_1(t)$ converge to the same trajectory
and the same occurs for the other state variables of the master and slave system.

7.3.2 Master-slave synchronization by linear feedback

Linear feedback may also be used for master-slave synchronization by adopting
the scheme of Figure 7.6. In this scheme, the slave is designed as the observer
of the master. The observer in control systems is a device that provides an
estimation of the state vector from the knowledge of the model, the input,

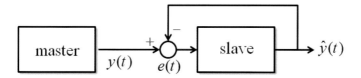

FIGURE 7.6
Master-slave synchronization based on linear feedback.

and output of the system [53]. The output of the master is indicated as $y(t)$, that of the slave (estimated output) as $\hat{y}(t)$.

According to this scheme, given a master system governed by the equations:

$$\dot{\mathbf{x}} = f(\mathbf{x})$$
$$y = \mathbf{C}\mathbf{x} \tag{7.12}$$

the slave system is designed as:

$$\dot{\hat{\mathbf{x}}} = f(\hat{\mathbf{x}}) + \mathbf{K}\mathbf{e}$$
$$\hat{y} = \mathbf{C}\hat{\mathbf{x}} \tag{7.13}$$

with $\mathbf{e}(t) = \mathbf{x}(t) - \hat{\mathbf{x}}(t)$ being the error between master and slave.

Synchronization, that is the condition for $\lim_{t\to\infty} \|\mathbf{x}(t) - \hat{\mathbf{x}}(t)\| = 0$, is achieved if the error dynamics

$$\dot{\mathbf{e}} = F(\mathbf{e}) \tag{7.14}$$

is designed so that the origin is a stable equilibrium point. The design parameter is the vector of gains K. We note that the observer-based approach can lead to a weak action on the slave to estimate the master state variables.

The problem is to study the best configuration that allows us to find a suitable observer scheme for which a Lyapunov function of the error dynamics can be chosen. The approach is general and based on the well-known theory for nonlinear system observers.

Example 7.2 _____

We illustrate the linear feedback method for master-slave synchronization of two Chua's circuits. We refer to the dimensionless equations of the circuit and suppose that the output variable in Equations (7.12) is the variable x of the Chua's circuit in Equations (5.5), that is we consider $\mathbf{C} = \begin{bmatrix} 1 & 0 & 0 \end{bmatrix}$. This yields the following equations for the master and slave system:

$$\dot{x} = \alpha[y - h(x)]$$
$$\dot{y} = x - y + z$$
$$\dot{z} = -\beta y$$
$$\dot{\hat{x}} = \alpha[\hat{y} - h(\hat{x})] + k_1(x - \hat{x}) \qquad (7.15)$$
$$\dot{\hat{y}} = \hat{x} - \hat{y} + \hat{z}$$
$$\dot{\hat{z}} = -\beta \hat{y}$$

These equations are implemented in the file `mslinearfeedbackchuaeqs.m` as follows:

```
function dxdt = mslinearfeedbackchuaeqs(t,x,k)

alpha = 9;
beta = 14.286;
c1=1/16;
c0=-1/6;

dxdt = [alpha*(x(2) - c1*x(1)^3-c0*x(1));
    x(1) - x(2) + x(3);
    - beta*x(2)
    alpha*(x(5) - c1*x(4)^3-c0*x(4))+k(1)*(x(1)-x(4));
    x(4) - x(5) + x(6)+k(2)*(x(1)-x(4));
    - beta*x(5)+k(3)*(x(1)-x(4))];
```

Note that the two systems have the same parameters α, β, c_0, and c_1. The vector of the observer gains, that is k $= \begin{bmatrix} k_1 & k_2 & k_3 \end{bmatrix}$, is passed as an external parameter of the function. We select k $= \begin{bmatrix} 1 & 0 & 0 \end{bmatrix}$ and, by integrating the equations, show that such values are suitable to attain synchronization:

```
[t,x]=ode45(@mslinearfeedbackchuaeqs,[0 200],[0.1 -0.05 0.3 0.3 0.2
    -0.4],'',[1 1 0]);
figure,plot(t,x(:,1:3:end)), xlabel('t'), ylabel('x(t)')
figure,plot(t,x(:,1)-x(:,4)), xlabel('t'), ylabel('x_1(t)-x_4(t)')
```

As shown in Figure 7.7, illustrating the trend of the MATLAB® variables $x_1(t)$ (corresponding to $x(t)$) and $x_2(t)$ (corresponding to $\hat{x}(t)$) and their difference, the two systems synchronize in a short time. The two variables $x(t)$ and $\hat{x}(t)$ are almost indistinguishable. The same holds for the other state variables of the two systems.

7.3.3   Master-slave synchronization by inverse system

The method of master-slave synchronization by design of the inverse system is strictly connected to the possibility of using chaos for analog cryptography [87]. Therefore, to illustrate it, we refer to a chaotic system driven by an external signal $s(t)$, that is the message to be sent in a secret way. Instead of an autonomous chaotic system, we consider a chaotic system which subject to $s(t)$ produces an output $y(t)$ that is a chaotic signal and that represents the signal transmitted as schematically illustrated in Figure 7.8.

The equations of the master system are:

$$\dot{\mathbf{x}} = f(\mathbf{x}, s)$$
$$y = g(\mathbf{x}, s) \qquad (7.16)$$

and those of the slave system:

$$\dot{\hat{\mathbf{x}}} = f(\hat{\mathbf{x}}, y)$$
$$\hat{y} = g(\hat{\mathbf{x}}, y) \qquad (7.17)$$

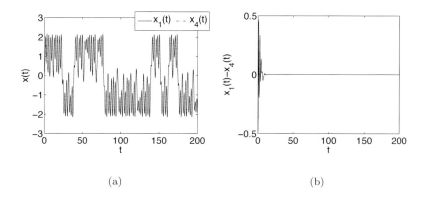

(a)   (b)

FIGURE 7.7
Master-slave synchronization via linear feedback of two Chua's circuits: (a) trends of $x_1(t)$ and $x_4(t)$ (these MATLAB® variables correspond to $x(t)$ and $\hat{x}(t)$ in Equations (7.15)); (b) difference $x_1(t) - x_4(t)$.

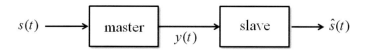

FIGURE 7.8
Message transmission through chaotic encryption.

The master system is chaotic when $s(t) = 0$. The presence of a non-zero signal $s(t)$ does not change the nature of the dynamics that remains chaotic. In fact, $s(t)$ is a weak signal. Still this signal is able to modify the dynamical evolution of the output signal. Another important point is that $s(t)$ is a low-level signal, but in any case not comparable to noise. With a proper design of the synchronization scheme asymptotically one has $\hat{\mathbf{x}} \to \mathbf{x}$, so that $\hat{s}(t) \simeq s(t)$, that is the message is recovered at the receiver, while it is masked during transmission as $y(t)$ is chaotic.

We illustrate the method for synchronization with reference to the Chua's circuit. The method is based on finding the inverse system, a problem which is far from trivial from a mathematical point of view. It turns out, however, that, when circuits are dealt with, the problem finds specific instances where it is much simplified.

The inverse system scheme for the Chua's circuit is illustrated in Figure 7.9. $i(t)$ is the signal to be transmitted and is injected in the circuit as

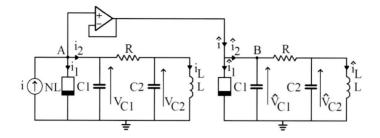

FIGURE 7.9
Master-slave synchronization based on the inverse system.

an independent current source. The voltage v_A from the master circuit (the Chua's circuit on the left) is sent to the slave circuit (the Chua's circuit on the right); the scheme imposes that $v_A \simeq v_B$ and, therefore, $\hat{i}_1(t) \to i_1(t)$ and $\hat{i}_2(t) \to i_2(t)$. From which it derives that $\hat{i}(t) \to i(t)$, thus measuring $\hat{i}(t)$ the message may be recovered.

This scheme is also referred to as *chaotic masking* since the message is added to a chaotic signal in the master. The aim of the slave is to synchronize with the master so that with a simple difference operation the message can be recovered. This is only possible if the slave has the same circuit parameters as the master: the circuit parameters thus represent the encryption key. Moreover, the initial conditions that are different in the master and in the slave make synchronization the crucial point of the encryption.

It is important to remark that only one scalar signal is transmitted, otherwise, transmitting more signals the problem becomes trivial. Even if analog cryptography cannot be used in digital communications where well-established methods for digital cryptography for security protocols exist, the use of chaos in cryptography is a useful idea for industrial applications where it is necessary to encode the signals coming from the parts of the system in a different way so that they can be distinguished by the others if appropriate keys are retained.

An example, discussed in [30], is in robot control. In robotics, ultrasonic devices are widely used as exteroceptive sensors for ranging measurements. In fact, these applications involve a large number of sonars operating concurrently, giving rise to the phenomenon of crosstalk, as each robot may be equipped with many sensors. Since chaotic systems have the property of sharp autocorrelation and uncorrelation between signals coming from different systems as well as signals coming from different attractors of the same system, chaos may be used to provide a unique signature for the source of information. To this aim, each sonar is driven with a suitable chaotic signal (in particular, the sequence of pulses emitted by each sonar is modulated by a chaotic signal so that the duration of the time interval between a pulse and the next one is regulated by a chaotic law) and, then, a matched filter technique is applied for a robust rejection of crosstalk and noise. The circuitry to realize this is simple as chaotic signals are generated with a low-cost analog circuitry.

Master-slave synchronization with the inverse system is further discussed in the part of the book dealing with laboratory experiments (Section 8.11), where the application for chaos encryption is illustrated.

7.4 Synchronization via diffusive coupling

In the schemes of the previous section the coupling between the systems was characterized by a specific direction, set in accordance to the flow of information, from the master circuit to the slave one. However, synchronization is also possible, and commonly found, when the interaction is mutual. In particular, this occurs each time the coupling is diffusive. The first example of diffusive coupling is a resistance placed between two capacitors representing the storage element associated to the state variables of two nonlinear circuits. As the current may flow in both directions, the interaction is bidirectional. Very often if the coupling is strong enough this results in the possibility of synchronizing the two circuits. The next example illustrates this phenomenon for a pair of Chua's circuits.

Example 7.3 ⎯⎯⎯⎯⎯⎯⎯⎯⎯⎯⎯⎯⎯⎯⎯⎯⎯⎯⎯⎯⎯⎯⎯⎯⎯⎯⎯⎯⎯⎯⎯⎯

Consider two Chua's circuits coupled by means of a resistor, R_c, connected between two corresponding capacitors as in Figure 7.10. In terms of dimensionless equations the coupling is described by the following equations:

$$
\begin{aligned}
\dot{x} &= \alpha[y - h(x)] + k_x(x' - x) \\
\dot{y} &= x - y + z \\
\dot{z} &= -\beta y \\
\dot{x}' &= \alpha[y' - h(x')] + k_x(x - x') \\
\dot{y}' &= x' - y' + z' \\
\dot{z}' &= -\beta y'
\end{aligned}
\tag{7.18}
$$

where $k_x = \frac{\alpha R}{R_c}$. The coupling acts on the x variable. On the contrary, if a resistor is placed between the corresponding capacitors C_2, then coupling is realized through the y variable.

Suppose now to perform the following experiment by using a variable coupling resistor, e.g., a potentiometer. Start from a high value of this resistor. If the resistance is high, a very small current will flow in it and the mutual interaction between the circuits will be so small that they practically evolve in an independent way: for a high value of the coupling resistor, the two circuits are not synchronized. Decreasing the resistance, it is possible to increase the strength of the mutual interaction until synchronization is achieved. Therefore, in this configuration, synchronization is achieved for a sufficiently low value of R_c or, equivalently, for a sufficiently high value of the coupling parameter k_x.

Numerical simulations can illustrate the example. They can be carried out by writing Equations (7.18) in a file `coupledchuaeqs.m` as follows:

```
function dxdt = coupledchuaeqs(t,x,kx)
```

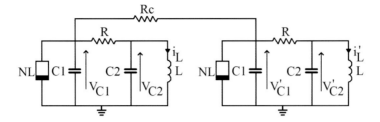

FIGURE 7.10
Synchronization through diffusive coupling between two Chua's circuits.

```
alpha = 9;
beta = 14.286;
c1=1/16;
c0=-1/6;

dxdt = [alpha*(x(2) - c1*x(1)^3-c0*x(1))+kx*(x(4)-x(1));
    x(1) - x(2) + x(3);
    - beta*x(2)
    alpha*(x(5) - c1*x(4)^3-c0*x(4))+kx*(x(1)-x(4));
    x(4) - x(5) + x(6);
    - beta*x(5)];
```

and then integrating them. Note that the coupling coefficient appears as a parameter, so that changing it is straightforward. Consider first zero coupling (i.e., $R_c \to \infty$):

```
[t,x]=ode45(@coupledchuaeqs,[0 200],[0.1 -0.05 0.3 0.3 0.2 -0.4],'',
0);
figure,plot(t,x(:,1:3:end)), xlabel('t'), ylabel('x(t)')
figure,plot(t,x(:,1)-x(:,4)), xlabel('t'), ylabel('x_1(t)-x_2(t)')
```

The trends of $x_1(t)$ and $x_2(t)$ reported in Figure 7.11(a) and the difference $x_1(t) - x_2(t)$ shown in Figure 7.11(b) are obtained. The systems starting from different initial conditions follow uncorrelated trajectories. Vice versa, when the coupling is high enough, for instance $k_x = 4$ as in the following:

```
[t,x]=ode45(@coupledchuaeqs,[0 200],[0.1 -0.05 0.3 0.3 0.2 -0.4],'',
4);
figure,plot(t,x(:,1:3:end)), xlabel('t'), ylabel('x(t)')
figure,plot(t,x(:,1)-x(:,4)), xlabel('t'), ylabel('x_1(t)-x_2(t)')
```

the two systems synchronize: the variables follow the same evolution (Figure 7.11(c)) and the error decreases in time (Figure 7.11(d)).

One of the most interesting features of diffusive coupling is that it straightforwardly generalizes to the case of more than two coupled units. In this form it is often found in nature. In addition, for such coupling there exists a widely used technique which enables the calculation of the range of the diffusive coupling coefficient for which the synchronous state is stable.

To illustrate this technique, known as the *master stability function (MSF) approach*, we will refer to a network of coupled dynamical units. The links are assumed to be time-invariant and the nodes having identical dynamics. We

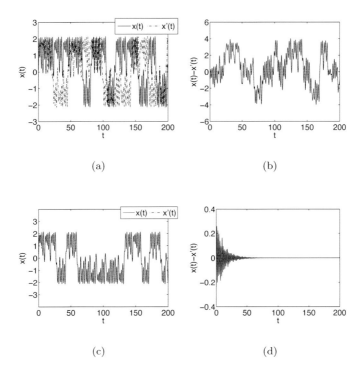

(a)   (b)

(c)   (d)

FIGURE 7.11
Synchronization via diffusive coupling of two Chua's circuits: (a) trends of
$x_1(t)$ and $x_2(t)$ for $k_x = 0$ (no synchronization); (b) difference $x_1(t) - x_2(t)$
for $k_x = 0$ (no synchronization); (c) trends of $x_1(t)$ and $x_2(t)$ for $k_x = 4$
(complete synchronization); (d) difference $x_1(t) - x_2(t)$ for $k_x = 4$ (complete
synchronization).

will refer only to bidirectional coupling, although the theory is more general.
We also note that several extensions of the basic approach presented here
have been proposed in the literature, for instance to deal with the case of
uncertainties in the node dynamics or of time-variant links, so that the state
of the art of the conditions where the MSF approach may be applied is larger
than that here discussed.

The reference model to introduce the MSF approach is the following

$$\dot{\mathbf{x}}^i = \mathbf{F}(\mathbf{x}^i) - K \sum_j g_{ij} \mathbf{H}(\mathbf{x}^j) \tag{7.19}$$

where $i = 1, \ldots, N$, $\mathbf{x}^i$ is an m-dimensional vector of dynamical variables of
the i-th node, $\dot{\mathbf{x}}^i = \mathbf{F}(\mathbf{x}^i)$ represents the dynamics of each isolated node, that

is the uncoupled dynamics, K is the coupling strength, $\mathbf{H} : \mathbb{R}^m \to \mathbb{R}^m$ is the coupling function and $G = [g_{ij}]$ is a zero-row sum $N \times N$ matrix modelling network connections. This matrix is known as the Laplacian of the network and along with the adjacency matrix is a widely used mathematical tool to describe a network. More precisely, the adjacency matrix $A = [A_{ij}]$ of a (for simplicity, unweighted and undirected) network is defined as $A_{ij} = 0$ if nodes i and j are not connected, and $A_{ij} = A_{ji} = 1$ if i and j are linked. In addition, $A_{ii} = 0$. Similarly, the network Laplacian is defined as $G_{ij} = 0$ if nodes i and j are not connected, $G_{ij} = -1$ if nodes i and j are linked, and $G_{ii} = k_i$, where k_i is the degree of node i, that is the number of links node i has ($k_i = \sum_j A_{ij}$). In matrix form the Laplacian is given by $G = D - A$, where D is a diagonal matrix containing the degrees of the network nodes.

The fact that the Laplacian is a zero-row sum matrix assures that the network configuration admits an invariant synchronization manifold, or equivalently that, when all the network nodes evolve along the same trajectory, the network is governed by the uncoupled dynamics at each identical node. In fact, the MSF approach derives the conditions under which N identical oscillators can be synchronized, under arbitrary network configurations provided that they admit an invariant synchronization manifold. We also note that diffusive coupling automatically satisfies this property, although clearly the field of applicability of the MSF approach is wider.

The original formulation of the MSF approach is due to Pecora and Carroll [69]. According to their analysis, the dynamics of the network is linearized around the synchronization manifold. Under the hypothesis that the network Laplacian is diagonalizable, a block diagonalized variational equation of the form

$$\dot{\xi}_h = [D\mathbf{F} - K\gamma_h D\mathbf{H}]\xi_h \qquad (7.20)$$

is obtained. It represents the dynamics of the system around the synchronization manifold, where γ_h is the h-th eigenvalue of G, $h = 1, \ldots, N$. $D\mathbf{F}$ and $D\mathbf{H}$ are the Jacobian matrices of $\mathbf{F}$ and $\mathbf{H}$ computed around the synchronous state, and are the same for each block. Therefore, the blocks of the diagonalized variational equation differ from each other only for the term $K\gamma_h$. If one wants to study synchronization properties with respect to different topologies, the variational equation must be studied as a function of a generic (complex) eigenvalue $\alpha + j\beta$. If the network is symmetrical as in the case of diffusive coupling, then the eigenvalues are real and one can further simplify the analysis by studying the variational equation as a function of a generic real eigenvalue α. Focusing on this latter case, the following master stability equation (MSE) is defined:

$$\dot{\zeta} = [D\mathbf{F} - \alpha D\mathbf{H}]\zeta. \qquad (7.21)$$

The maximum (conditional) Lyapunov exponent λ_{max} of the MSE is studied as a function of α, thus obtaining the MSF, i.e., $\lambda_{max} = \lambda_{max}(\alpha)$. Then,

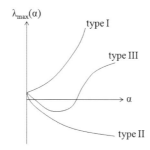

FIGURE 7.12
Different classes of master stability functions for coupled chaotic systems.

the stability of the synchronization manifold in a given network can be evaluated by computing the eigenvalues γ_h (with $h = 2, \ldots, N$) of the matrix G and studying the sign of λ_{max} at the points $\alpha = K\gamma_h$. If all eigenmodes with $h = 2, \ldots, N$ are stable, then the synchronous state is stable at the given coupling strength.

The functional dependence of λ_{max} on α gives rise to three different cases [6], shown in Figure 7.12. The first case (type I) is such that λ_{max} is positive $\forall \alpha$ and, thus, the network nodes cannot be synchronized. In the second case (type II) λ_{max} assumes negative values above a threshold value, and one always gets a stable synchronous state for high enough coupling strength. In the third case (type III) λ_{max} is negative only in a closed interval of values of α, that is, network nodes can be synchronized only if $K\gamma_h$ for $\forall h = 2, \ldots, N$ lies in this interval.

The MSF approach gives a necessary condition (the negativity of all Lyapunov exponents transverse to the synchronization manifold) for the stability of complete synchronization. The method can be applied to arbitrary either weighted or unweighted topologies. As previously noted, the MSF approach here presented for diagonalizable networks (an assumption always valid for undirected networks, where the Laplacian is symmetric) has been generalized to non-diagonalizable networks, to scenarios where the nodes are not identical and to topologies with time-varying links.

Example 7.4 _____

To illustrate the MSF approach, we consider a network of $N = 10$ Rössler systems modelled by the following equations:

$$
\begin{aligned}
\dot{x}^i &= -y^i - z^i - \sigma \sum_{h=1}^{N} g_{ih} x^h \\
\dot{y}^i &= x^i + a y^i \\
\dot{z}^i &= b + (x^i - c) z^i
\end{aligned}
\tag{7.22}
$$

where all the units are identical and chaotic ($a = b = 0.2$ and $c = 9$). g_{ih} represent

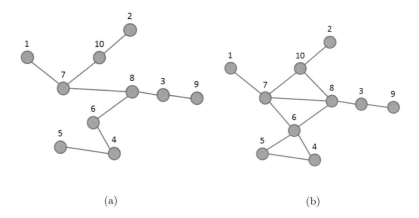

(a)   (b)

FIGURE 7.13
Networks of Rössler oscillators: (a) a network that cannot be synchronized;
(b) a network that can be synchronized for a proper choice of the coupling
coefficient.

the coefficients of the Laplacian G of the network and σ the coupling coefficient.
Synchronizability of the network can be studied by inspecting the MSF of the
Rössler systems when coupled through the variable x on the dynamics of the same
variable. For many chaotic systems and types of coupling the MSF is reported in
the literature. One can, for instance, refer to paper [40], which shows that for the
case under investigation the MSF is of type III with $\alpha_1 \simeq 0.186$ and $\alpha_2 \simeq 4.614$.
To check if a given network is synchronizable, we have thus to check the conditions:

$$\sigma\gamma_2 > \alpha_1$$
$$\sigma\gamma_N < \alpha_2 \qquad (7.23)$$

which assures that all the eigenvalues are in the stability region. Consider the network in Figure 7.13(a). For this network $\gamma_2 = 0.1859$ and $\gamma_N = 4.7517$. Therefore,
since $\gamma_N/\gamma_2 > \alpha_2/\alpha_1$, conditions (7.23) cannot be satisfied. The scenario is different for the network of Figure 7.13(b), obtained from that of Figure 7.13(a) with
the addition of few links. In this case, $\gamma_2 = 0.3657$ and $\gamma_N = 5.6736$, so that the
coupling coefficient has to satisfy

$$0.5086 = \frac{\alpha_1}{\gamma_2} = \sigma_1 \le \sigma \le \sigma_2 = \frac{\alpha_2}{\gamma_N} = 0.8132 \qquad (7.24)$$

We can for instance select $\sigma = 0.75$. To verify the presence of synchronization the
following commands may be used:

```
%network definition
N=10;
A=zeros(N,N);
links=[1 7; 2 10; 3 8; 3 9; 4 5; 4 6; 5 6; 6 7; 6 8; 7 8; 7 10;
8 10];
for l=1:length(links)
    A(links(l,1),links(l,2))=1;
    A(links(l,2),links(l,1))=1;
end
```

```
%network Laplacian
degree=sum(A,2);
Gmatrix=diag(degree)-A;

dt=0.001;
steps=500/dt;

%parameters of the units
ar=0.2;
br=0.2;
cr=9;

%coupling
sigma=0.75;

%initialization
x=zeros(steps,N);
y=zeros(steps,N);
z=zeros(steps,N);

%initial conditions for the oscillators
xold=(30*rand(N,1)-15)/5;
yold=(30*rand(N,1)-15)/5;
zold=(40*rand(N,1)-5)/5;

error=zeros(steps,1);

for t=1:steps

    coupling=-Gmatrix*xold;

    dxdt=-yold-zold+sigma*coupling;
    dydt=xold+ar*yold;
    dzdt=br+zold.*(xold-cr);

    xnew=xold+dt*dxdt;
    ynew=yold+dt*dydt;
    znew=zold+dt*dzdt;

    xold=xnew;
    yold=ynew;
    zold=znew;

    error(t)=mean(abs(xnew(2:N)-xnew(1))+abs(ynew(2:N)-ynew(1))
    +abs(znew(2:N)-znew(1)))/3;

    x(t,:)=xnew;
    y(t,:)=ynew;
    z(t,:)=znew;

end

figure, plot([1:steps]*dt,error,'k'), xlabel('t'), ylabel('e(t)')
```

The network equations are integrated with a Euler algorithm and step size equal to $dt = 0.001$. The overall synchronization error is computed as:

$$e(t) = \frac{1}{N-1} \sum_{h=2}^{N} \frac{|x^h(t) - x^1(t)| + |y^h(t) - y^1(t)| + |z^h(t) - z^1(t)|}{3} \qquad (7.25)$$

where node 1 has been arbitrarily chosen as reference (recall that the MSF approach predicts complete synchronization for all the network pairs, that is, $\mathbf{x}^1 = \mathbf{x}^2 = \ldots = \mathbf{x}^N$). The evolution of the error is shown in Figure 7.14.

To conclude the example, we report the MATLAB® commands that can be used to check if conditions (7.23) are satisfied:

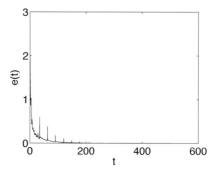

FIGURE 7.14
Synchronization error $e(t)$ as in Equation (7.25) for a network of $N = 10$ Rössler oscillators coupled as in Figure 7.13(b) with $\sigma = 0.75$.

```
alpha1=0.186;
alpha2=4.614;
N=10;
A=zeros(N,N);
links=[1 7; 2 10; 3 8; 3 9; 4 5; 4 6; 5 6; 6 7; 6 8; 7 8; 7 10;
 8 10];
for l=1:length(links)
    A(links(1,1),links(1,2))=1;
    A(links(1,2),links(1,1))=1;
end
degree=sum(A,2);
G=diag(degree)-A;
lg=eig(G);
sigma1=alpha1/lg(2)
sigma2=alpha2/lg(end)
```

7.5   Principles of chaos control

Chaos can be found in many natural systems and often it also exists in engineering processes and devices, so that the need of controlling it has arisen. The term chaos control in a broad sense includes different issues such as:

- to stabilize a limit cycle included in the attractor;

- to suppress chaos;

- to create chaos in systems where, without control, it is not found.

So, chaos control refers to the control of the transition from disorder to order as well as to that from order to disorder. In this sense, it is fundamental to ask whether the chaotic behavior is desired (and chaos control is used to elicit it) or undesired (and chaos control has to be used to suppress it). Although chaos control can be dealt with as a classical control problem of a nonlinear system, it requires an accurate knowledge of the system and the adopted strategy can change depending on the considered circuit or system to control.

For systems that benefit from the presence of chaos or for those that are intentionally designed to be chaotic (in secure communications for instance), maintaining chaos is fundamental. On the contrary, in other systems the presence of chaos should be absolutely avoided.

Another reason to investigate chaos control is that, given the high quantity of information contained in the strange attractor of a chaotic system, the possibility to encode this type of information is related to the stabilization of unstable limit cycles which requires suitable control strategies. Of course, there are control strategies that need a significant amount of energy and others that can be performed with low energy level, thus amplifying the effect of weak interactions and small control signals. Moreover, the various techniques must take into account that a strange chaotic attractor has embedded a dense set of unstable limit cycles very sensitive to the control action.

7.6   Strategies for chaos control

The strategies for chaos control that are mainly studied in literature are based on the following principles:

- adaptation of accessible system parameters like bifurcation parameters;

- modification of the system dynamics;

- injection of external signals;

- feedback techniques.

7.6.1   Adaptation of accessible system parameters

If a complete set of bifurcation diagrams is available, one can easily determine for a system which parameters to keep fixed and which (possibly limiting their number to one or few) should be taken as control parameters to achieve a specific goal. In fact, the bifurcation diagrams provide a complete view of the system dynamics and, therefore, based on them one can envisage how to drive the system towards the desired behavior through the appropriate parameters.

This technique is very intuitive and emphasizes the importance of the bifurcation diagrams in chaos control. Generally, the technique gives no guarantee that a small amount of energy is required to change the value of the control parameter.

7.6.2   Entrainment control

The entrainment control is an open-loop technique. Given $\dot{\mathbf{x}} = F(\mathbf{x}, t)$ the system to be controlled and said $\tilde{\mathbf{x}}(t)$ the target trajectory, the control problem consists of modifying the system equation so that

$$\lim_{t \to \infty} \|\mathbf{x}(t) - \tilde{\mathbf{x}}(t)\| = 0 \tag{7.26}$$

To this aim, the system equation is changed to include a further additive term $\tilde{G}(\dot{\tilde{\mathbf{x}}}, \tilde{\mathbf{x}})$ so that the new system

$$\dot{\mathbf{x}} = F(\mathbf{x}, t) + \tilde{G}(\dot{\tilde{\mathbf{x}}}, \tilde{\mathbf{x}}) \tag{7.27}$$

admits a stable solution $\tilde{\mathbf{x}}(t)$.

The peculiarities of this method are:

- it does not require feedback and, therefore, an accurate model of the system has to be used;

- the modification of the original system dynamics included in the method generally requires a strong action;

- a not large set of trajectories can be controlled.

7.6.3   Weak periodic perturbation

This method is based on the injection of periodic signals in the system. The amplitude and frequency of these signals are the control parameters. The interaction between the periodic perturbation and the chaotic system results in the desired trajectory.

The method is easy to apply and very efficient as it requires low-energy signals, in agreement with the concept of weak interactions. As a special case it includes the vibrational control technique. In fact, this latter is based on the use of small external perturbation to stabilize equilibrium points, limit cycles and so on.

The method is based on the fact that, given an autonomous system $\dot{\mathbf{x}} = F(\mathbf{x})$ to be controlled, the injection of the periodic control signal makes it non-autonomous:

$$\dot{\tilde{\mathbf{x}}} = \tilde{F}(\tilde{\mathbf{x}}, t) \tag{7.28}$$

Unfortunately, the method, although very efficient, has not a complete

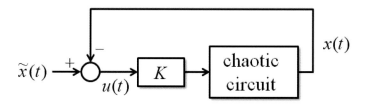

FIGURE 7.15
Linear feedback control for chaos control.

theory and often numerical simulations have to be used to study its reliability. In the experiment part of the book (Section 8.13) a practical example based on this control technique is presented.

7.6.4 Feedback control

When it is difficult to establish ad hoc strategies for chaos control, the classical feedback control is used. However, also in this case the control design technique must be studied for the chaos control constraints. The easier control technique in this sense achieves the best results, that is linear feedback control or PID control.

The linear feedback control is based on the scheme of Figure 7.15. Given $\tilde{\mathbf{x}}(t)$ a solution of a chaotic system, the control task is to stabilize this trajectory by using a linear feedback law $u(t) = \mathrm{K}(\tilde{\mathbf{x}}(t) - \mathbf{x}(t))$. Any solution of the original system can be considered in the control task. However, the difficulty of this method is that it needs the feedback interaction of many state variables [17].

7.6.5 The OGY approach

The main idea of the Ott–Grebogi–Yorke (OGY) approach is to consider one of the unstable periodic orbits embedded in a strange attractor of a chaotic system and slightly perturb it, so that the system is driven towards the desired orbit. Due to the fact that chaos is ergodic, at some point in time the solution will pass near the orbit, therefore allowing us to apply a linearization procedure.

Let us consider a continuous-time chaotic system and a Poincaré map, where each point represents an (unstable) limit cycle for the system and an (unstable) equilibrium point for the map. The idea of the OGY method is to build the Poincaré map around a point by using linearization. Suppose that the dynamical behavior of the system depends on a parameter p, so that the Poincaré map is written as:

$$\mathbf{x}_{k+1} = f(\mathbf{x}_k, p) \tag{7.29}$$

A fixed point $\mathbf{x}_{k+1} = \mathbf{x}_k = \bar{\mathbf{x}}$ is associated to $p = \bar{p}$. Assuming that this point is unstable (in the corresponding continuous-time system this point represents an unstable limit cycle), we linearize around it and apply a control law to stabilize it. The linearized system is given by:

$$\mathbf{x}_{k+1} - \bar{\mathbf{x}} = \mathrm{A}(\mathbf{x}_k - \bar{\mathbf{x}}) + \mathrm{B}(p - \bar{p}) \tag{7.30}$$

where A and B are the Jacobian matrices:

$$\begin{aligned} \mathrm{A} &= \left.\frac{\partial f}{\partial \mathbf{x}}\right|_{(\bar{\mathbf{x}},\bar{p})} \\ \mathrm{B} &= \left.\frac{\partial f}{\partial p}\right|_{(\bar{\mathbf{x}},\bar{p})} \end{aligned} \tag{7.31}$$

To stabilize the equilibrium point $\bar{\mathbf{x}}$, the following control law is applied to system (7.30):

$$p - \bar{p} = -\mathrm{K}(\mathbf{x}_k - \bar{\mathbf{x}}) \tag{7.32}$$

This yields

$$\mathbf{x}_{k+1} - \bar{\mathbf{x}} = (\mathrm{A} - \mathrm{BK})(\mathbf{x}_k - \bar{\mathbf{x}}) \tag{7.33}$$

Therefore, K have to be chosen so that the closed loop matrix $\mathrm{A} - \mathrm{BK}$ has all eigenvalues inside the unit circle. In this way, the equilibrium point $\bar{\mathbf{x}}$ for the Poincaré map and the corresponding limit cycle are stabilized. The difficulty of the control technique lies in the estimation of the matrices A and B from the data through some identification procedure.

In summary, the OGY method consists of the following steps:

- define an embedded system from the experimental data;

- build the Poincaré map of the embedded system and identify the discrete map to which apply the fixed point stabilization;

- linearize the Poincaré map and find the matrices A and B;

- find the gains K and implement the OGY control law.

The OGY technique is a feedback technique that does not require the model of the dynamics. It is sensitive to noise as all the techniques that require identification from data.

Example 7.5 _____

As an example we apply the OGY technique to control chaos in the logistic map. More in particular, let us consider the logistic map with $\bar{r} = 3.8$. For this value the logistic map is chaotic. However, the chaotic set embeds several unstable periodic orbits and an unstable fixed point $\bar{x} \simeq 0.7368$. Suppose that we want to stabilize this equilibrium point. For this purpose we use the OGY control.

Let us first calculate the control law (7.32) that for convenience is rewritten as

$$\Delta r = r - \bar{r} = -\mathrm{K}(x_k - \bar{x}) \tag{7.34}$$

It is applied to the logistic map in this way:

$$x_{k+1} = \bar{r}x_k(1 - x_k) + u_k \tag{7.35}$$

with $u_k = \Delta r x_k(1 - x_k)$.

The gain K is evaluated so that $A - BK$ has an eigenvalue inside the unit circle. We can fix it in the origin, that is we select $A - BK = 0$ and thus $K = A/B$ where A and B are calculated according to Equations (7.31):

$$\begin{aligned} A &= \left.\frac{\partial f}{\partial x}\right|_{(\bar{x},\bar{r})} = \bar{r}(1 - 2\bar{x}) \\ B &= \left.\frac{\partial f}{\partial r}\right|_{(\bar{x},\bar{r})} = \bar{x}(1 - \bar{x}) \end{aligned} \tag{7.36}$$

Therefore, one gets:

$$K = \frac{\bar{r}(1 - 2\bar{x})}{\bar{x}(1 - \bar{x})} \tag{7.37}$$

The following MATLAB® commands may be used to verify the effectiveness of the control.

```
xbar=0.7368;
rbar=3.8;
Nsteps=1000;
x=zeros(Nsteps,1);
u=zeros(Nsteps,1);
x(1)=0.23;
k=rbar*(1-2*xbar)/(xbar*(1-xbar));
for i=2:Nsteps
    deltar=-k*(x(i-1)-xbar);
    if abs(x(i-1)-xbar)<0.002, u(i)=deltar*x(i-1)*(1-x(i-1)); end
    x(i)=rbar*x(i-1)*(1-x(i-1))+u(i);
end

figure,plot(x,'k.')
xlabel('k')
ylabel('x_k')
```

Figure 7.16 shows the trend of x_k. The control becomes active when the ergodic trajectory passes close to $\bar{x}$ (the threshold is set to 0.002), after that the equilibrium point is stabilized with a small control action. Note that the time to activate the control depends on the initial condition (the reader can change the value of it and check that the trajectory will approach $\bar{x}$ in shorter or longer times depending on the specific initial condition).

7.6.6   Noise for chaos control

In linear systems the effect of the presence of noise is usually detrimental and scientists and engineers try to avoid it. On the contrary, in some nonlinear systems noise can positively influence the system behavior. Our attention here is focused on an example where noise can be used to control chaos.

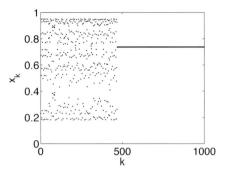

FIGURE 7.16
OGY control of the logistic map to stabilize an unstable equilibrium point of
the map.

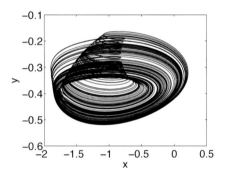

FIGURE 7.17
Attractor of the forced van der Pol oscillator (7.38).

Example 7.6 _____

Consider the forced van der Pol oscillator

$$\begin{aligned}\dot{x} &= x - \tfrac{x^3}{3} - y + A\cos(\omega t)\\ \dot{y} &= c(x + a - by)\end{aligned} \tag{7.38}$$

This system exhibits a period doubling route to chaos and for the parameter values
fixed as $\omega = 1$, $A = 0.74$, $a = 0.7$, $b = 0.8$, $c = 0.1$ a chaotic attractor is observed.
Figure 7.17 illustrates this chaotic attractor.
Consider now the introduction of a noise term in the equations of the oscillators:

$$\begin{aligned}\dot{x} &= x - \tfrac{x^3}{3} - y + A\cos(\omega t) + \eta(t)\\ \dot{y} &= c(x + a - by)\end{aligned} \tag{7.39}$$

where $\eta(t)$ represents a zero-mean Gaussian noise with standard deviation σ. Depending on the value of the standard deviation the presence of this stochastic input
can suppress chaos: in particular, this occurs if $\sigma \geq 0.03$ [17]. To verify this finding,

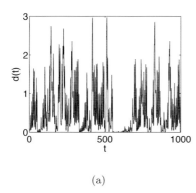 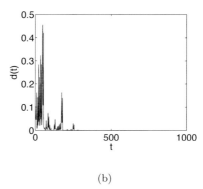

(a) (b)

FIGURE 7.18

Chaos suppression by noise in the forced van der Pol oscillator. Evolution of $d(t)$ for (a) $\sigma = 0.003$ and (b) $\sigma = 0.03$.

we consider trajectories starting from two nearby initial conditions and monitor the difference $d(t) = x(t) - x'(t)$, where $x(t)$ and $x'(t)$ are the evolutions of the state variables obtained starting from the two initial conditions. We note that $d(t)$ is irregular when the noise level is low (Figure 7.18(a)), while $d(t) \to 0$ for $\sigma \geq 0.03$ (Figure 7.18(b)), meaning that the system is no more sensitive to initial conditions. This result clearly shows that the noise can be used to tame chaos.

We now briefly illustrate the MATLAB® commands that can be used to reproduce the experiment discussed. First, note that system (7.39) is described by a *stochastic differential equation*. To integrate this, the Euler–Maruyama method, which is an extension of the Euler integration algorithm to stochastic differential equations, can be used. According to this method, each sample of the solution of a stochastic equation has to be computed as $x_{k+1} = x_k + f(x_k)\Delta t + \sigma \eta_k$. The method is implemented for the forced van der Pol oscillator (7.39) with the following commands, here reported with $\sigma = 0.03$:

```
omega=1;
A=0.74;
a=0.7;
b=0.8;
c=0.1;

dt=0.1;

steps=1000/dt;
x=zeros(steps,1);
y=zeros(steps,1);
x(1)=-1;
y(1)=-0.2;

x2=zeros(steps,1);
y2=zeros(steps,1);
x2(1)=-0.9;
y2(1)=-0.15;

n=0;
sigma=0.03;
```

```
for t=2:steps

    n=sigma*randn;

    dxdt=x(t-1)-x(t-1).^3/3-y(t-1)+A*cos(omega*(t-1)*dt);
    dydt=c*(x(t-1)+a-b*y(t-1));

    x(t)=x(t-1)+dt*dxdt+n;
    y(t)=y(t-1)+dt*dydt;


    dxdt2=x2(t-1)-x2(t-1).^3/3-y2(t-1)+A*cos(omega*(t-1)*dt);
    dydt2=c*(x2(t-1)+a-b*y2(t-1));

    x2(t)=x2(t-1)+dt*dxdt2+n;
    y2(t)=y2(t-1)+dt*dydt2;


end

figure,plot([1:steps]*dt,sqrt((x-x2).^2+(y-y2).^2),'k')
xlabel('t'), ylabel('d(t)')
```

7.7 Uncertain large-scale nonlinear circuits: spatio-temporal chaos control

In this section we study one of the extreme regions of the modified Varela diagram reported in Figure 1.3, namely the case of large-scale systems with high degree of nonlinearity and uncertainty. The discussion mainly relies on the paradigmatic results reported in [7] regarding arrays of coupled nonlinear pendula, but the use of uncertainty in order to tame chaos is a more general phenomenon, valid also for other dynamical systems.

The model consists of a large number of dynamical systems connected by a diffusive coupling network which acts locally, i.e., dynamical systems are arranged in a chain, in which each element is coupled with the two proximal elements only. This framework has been originally discussed in [7], where each dynamical system represents the mathematical model of a driven damped pendulum.

Let us consider an array of $N = 128$ driven damped pendula, each governed by the following equation:

$$ml_i^2 \ddot{\theta}_i + \gamma \dot{\theta}_i = -mgl_i \sin\theta_i + \tau' + \tau \sin\omega t + K(\theta_{i+1} + \theta_{i-1} - 2\theta_i) \quad (7.40)$$

where θ_i is the phase of the i-th pendulum, l_i is its length, $m = 1$ is the mass of the pendulum bob, $g = 1$ is the gravity, $\gamma = 0.75$ is the damping, $\tau' = 0.7155$ and $\tau = 0.4$ are the continuous and alternating components of the torque, ω is the angular frequency, and K is the coupling. The ends of the chain are considered free to move, which correspond to set zero-flux boundary conditions, i.e., $\theta_0 = \theta_1$ and $\theta_{N+1} = \theta_N$.

At first, we focus on the case in which $l_i = 1\ \forall i = 1\ldots N$, i.e., the uncertainty is neglected and all pendula have the same parameters. In these conditions, when $K = 0$ (uncoupled pendula), each element undergoes a chaotic unsynchronized oscillation.

A weak coupling, included by varying the coefficient to $K = 0.5$, is not able to drive the system towards a complete synchronization, as it can be seen from the inspection of Figure 7.19(a) where the evolution of the angular speed of each pundulum is reported. The figure has been obtained by defining the system with the following MATLAB® commands:

```
function D = oscind(t,xx)
 N=128;
 A=0.4;
 A1=0.7155;
 OD=0.25;
 K=0.5;
 G=0.75;
 O=1;
 for i=1:N;
     x(1,i)=xx(2*i-1);
     x(2,i)=xx(2*i);
     x(3,i)=xx(2*N+i);
 end;
 for i=1:N;
     G=G/x(3,i)^2;
        O=O/x(3,i);
           A1=A1/x(3,i)^2;
           A=A/x(3,i)^2;
           G=G/x(3,i)^2;
           R=R/x(3,1)^2;
     if i==1;
        D(2*i-1,1)=x(2,i);
        D(2*i,1)=(-G*x(2,i)-O*sin(x(1,i))+A1+A*sin(OD*t)
        +K*(x(1,i+1)-x(1,i)));
        D(2*N+i,1)=0;
     elseif i==N;
        D(2*i-1,1)=x(2,i);
        D(2*i,1)=(-G*x(2,i)-O*sin(x(1,i))+A1+A*sin(OD*t)
        +K*(x(1,i-1)-x(1,i)));
        D(2*N+i,1)=0;
     else
        D(2*i-1,1)=x(2,i);
        D(2*i,1)=(-G*x(2,i)-O*sin(x(1,i))+A1+A*sin(OD*t)
        +K*(x(1,i+1)+x(1,i-1)-2*x(1,i)));
        D(2*N+i,1)=0;
     end
 end
end
```

and then integrating the system equations with the command:

```
[T,Y]=ode45(@oscind,[0:0.01:500],[rand(1,128*2) ones(1,128)]);
```

In this scenario, the introduction of uncertainty on the system parameters plays a nontrivial role. Consider now an uncertainty acting on pendulum length, i.e., consider that l_i are stochastic variables drawn from a uniform distribution with values in the interval $[0.8, 1.2]$, thus introducing a $\pm 20\%$ uncertainty on l_i. The result of considering uncertain parameters is to drive the system towards a higher degree of coordination. The pendula are now subjected to synchronized angular speeds despite the difference in their length, as can be noticed in Figure 7.19(b), which has been realized using the command

```
[T,Y]=ode45(@oscind,[0:0.01:500],[rand(1,128*2) 0.8+0.4*rand(1,128)]);
```

Furthermore, it is interesting to note that the beneficial effect of uncertainty on the global coordination of the system is sensibly related to the specific level of uncertainty considered. When considering an uncertainty of the 10% applied to the same parameter, the global behavior is shown in Figure 7.19(c), where a lower degree of synchronization is observed.

The effect of uncertainty on large-scale systems of nonlinear coupled units is therefore nontrivial, and can be often beneficial in terms of the coordination of the global behavior. This concept is ubiquitous in nonlinear systems and similar effects can be retrieved for different dynamics, such as the chaotic Duffing oscillator [7] and the Chua's circuit. For this latter case, uncertainty leads to the coexistence of both synchronous and unsynchronous chaotic oscillations in the same structure, as well as to the emergence of multistable attractors. The reader can adapt the reported commands to the specific case study modifying the dynamical equations. Boundary conditions can also be varied acting on the coupling term of the first and last element of the array.

7.8 General remarks on chaos control

The topic of chaos control is vast and intriguing. The reason is that, even if chaos control can be faced with the classical feedback control theory, the peculiarities of chaos, the particular problems that arise in terms of unstable trajectories to be stabilized, the intensity of the control law and the appropriate performance that needs to be achieved, make the theme of extreme interest both as regards the study of new methods and the deep understanding of the main characteristics of chaos control. Moreover, chaos control is related to a lot of real-world problems, first of all, in biology and medicine. Indeed, one can take inspiration from biological feedback schemes and strategies to understand the mechanisms that allow us to generate stable and precise regulations.

In this chapter only a few methods have been discussed; most of them are related with systems with few state variables. The control techniques for systems with a high number of state variables is a still open field of research. The main guidelines that we have outlined are essential to understand the problem and to introduce the reader to the experiments on chaos taming reported in the second part of the book.

In the case of chaos control in CNNs and in distributed systems in general, uncertainty and noise also play a role that has to be considered to define new control paradigms, which, however, go beyond the scope of the book. We have only presented an example where uncertainty plays a fundamental role in achieving the global behavior of the system.

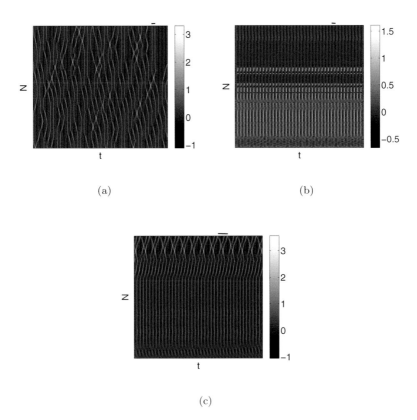

FIGURE 7.19

Spatio-temporal evolution of the angular speed of $N = 128$ coupled pendula:
(a) unsynchronized behavior of identical pendula, (b) regularized behavior
in presence of an uncertainty of 20% on pendula length, and (c) regions of
unsynchronized motion in the presence of an uncertainty of the 10% on pendula length. The horizontal bars in the panels represent the length of periodic
spatio-temporal patterns. Parameters are as fixed in the text, angular speeds
are color coded as in the colorbar.

We also note that, while we have focused on control strategies which are performed by external devices, many natural systems are self-organized for the emergence of order and control. Kauffman in his book, *The Origins of Order* [45], points out that "simple and complex systems exhibit the spontaneous emergence of order, the occurrence of self-organization." This means that many systems contain internal mechanisms for their regulation and control without the need for external control units.

Another important topic, which has not been faced in this chapter, is chaos transient and intermittency control. This new appealing topic requires rethinking and rearranging the concepts from chaos control and from bifurcation theory for better understanding of the emergent strategies for addressing it.

7.9   Exercises

1. Apply the method of master-slave synchronization based on system decomposition on two chaotic Chua's circuits.

2. Consider two hyperchaotic circuits and synchronize them by using one of the schemes presented in the chapter.

3. Consider three Lorenz systems working in chaotic conditions and diffusively coupled into a ring configuration. Find the synchronization conditions in terms of the coupling coefficient.

4. Consider a network of $N = 10$ chaotic Duffing systems coupled with diffusive coupling. Write a MATLAB® procedure to study network synchronization with respect to different network topologies.

5. Introduce in a node of the previous network a Gaussian noise with variance σ and discuss the system behavior.

6. Consider a network of Rössler systems as in Equations (7.22) and fix a star topology. Study synchronization with respect to the number N of oscillators.

7. Repeat the previous exercise for a ring topology.

8. Consider a chaotic system with an MSF of type III with $\alpha_1 = 0.5$ and $\alpha_2 = 4$. Report two examples of networks of such units, one that can be synchronized and one that cannot.

9. Consider the Duffing system

$$\ddot{x} + \gamma\dot{x} + x^3 = q\cos(\omega t) \tag{7.41}$$

with $\gamma = 0.3$, $\omega = 1$ and $q = 8.85$ such that the behavior is chaotic. Find a weak periodic perturbation able to suppress chaos.

10. Extract some unstable trajectories of the double scroll Chua's attractor. First, make some simulations by perturbing the bifurcation parameters and looking for some limit cycles. Then, consider such orbits as reference trajectories for the double scroll Chua's attractor and try to stabilize them by using linear feedback control.

Further reading

For additional information on the topics of the chapter, the following references may be consulted: [5], [14], [17], [18], [28], [45], [51], [61], [65], [71], [75], [83], [87].

8

Experiments and applications

CONTENTS

In this chapter a gallery of simple and significant experiments showing the behavior of some interesting nonlinear circuits is presented with the fundamental aim of proposing practical experiences that the reader can directly realize being stimulated to observe and characterize different behaviors and to invent new experiments.

The instrumentation [94] required to realize the experiments proposed in this chapter consists of the following equipment:

1. analog or digital multimeter;

2. dual-voltage power supply;

3. waveform/function generator;

4. oscilloscope.

Some of the instrumentation may be replaced by simple and low-cost solutions:

- the dual-voltage power supply can be obtained connecting in series two 9V batteries;

- the oscilloscopes and waveform/function generators can be replaced by a standard PC audio board through the line-in/line-out plugs by using free software available on the internet, e.g., [77].

The electrical components used are:

- wires;

- resistors;

- capacitors;

- inductors;

- diodes/LEDs;

- transistors;

- standard off-the-shelf operational amplifiers (e.g., TL084);

- standard off-the-shelf analog multipliers (e.g., AD633);

- breadboard to connect components;

- Arduino® board.

Components can be taken also from recycled electronic material. The authors invite the reader to prefer recycled components, which are also easily available from general purpose dismissed electronic devices.

The various experiments proposed in this chapter have been chosen to mirror the guidelines followed by the book. Furthermore, the aim of the practical experiments collected in this chapter is not limited to proving the simplicity of the implementation and of the study of complex dynamics by means of nonlinear circuits, but also at showing some of their useful applications.

Most of the experiments are designed to emphasize the role of elementary cells, such as those introduced in CNNs, when coupled in large ensembles. Spatio-temporal behavior induced by the mechanism of coupling among circuits is discussed and tested.

The added value given by nonlinear circuits with respect to the linear ones in many applications can be demonstrated with simple experiments. The underlying idea is to explain, study, and characterize the behavior of dynamical systems in terms of low-cost electronic circuits and equipment [12]. Nonlinear technology is indeed successful and useful and the experiments proposed in this chapter can convince the reader of this. The immediate and impressive results that can be obtained by using this simple electronic experimental approach will encourage readers to make more and more experiments.

8.1 Hewlett oscillator

The first experiment deals with the Hewlett oscillator discussed in Section 4.8 and reported in Figure 8.1. The oscillation frequency of this circuit is given by

$$f = \frac{1}{2\pi RC}$$

and, therefore, it is controllable by changing the values of R and C. The circuit originates from the passive network reported in Figure 8.2, where $G(s) = \frac{V_{out}}{V_{in}} = \frac{sRC}{(sRC+1)^2+sRC}$.

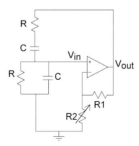

FIGURE 8.1
Implementation of the Hewlett oscillator.

Let us compute $G(j\omega)$ for $\omega = 2\pi f = \frac{1}{RC}$. It results in $G(j\omega) = \frac{1}{3}$; this means that $V_{out} = \frac{1}{3}V_{in}$ and that V_{out} is in phase when the input signal V_{in}

has a frequency of $f = \frac{1}{2\pi RC}$. When the operational amplifier is introduced in the circuit, as in Figure 8.1, the output is fed back both to the positive input and to the negative input. Note that V_{in} and V_{out} are exchanged with respect to the passive network. Therefore, the Barkausen phase condition for oscillation is satisfied since the output and input signals in the amplifier are both in phase. For the second Barkausen condition, regarding the gain, from simple circuit considerations, we derive that $(1+\frac{R_1}{R_2}) > 3$. Therefore, to achieve this condition in the circuit the value of the potentiometer implementing R_2 must be varied so that $R_1 > 2R_2$.

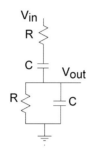

FIGURE 8.2
Hewlett oscillator passive network.

The circuit, shown in Figure 8.3(a) is powered by using two batteries connected in series to provide a dual voltage power supply with $V_{sup} = \pm 9V$: the reference ground is the central node of the series, while the positive and negative voltages are on the positive and negative battery poles, respectively.

The countereffect of increasing the gain of the amplifier is the presence of distortion effects as shown in the sequence of Figures 8.3(b)-(d) illustrating some experimental results for different values of the gain.

8.2   Van der Pol oscillator

The electronic scheme of the van der Pol oscillator is reported in Figure 8.4. In Figure 8.5(a) a circuit implementation is reported. In this circuit, analog multipliers are used. Each analog multiplier produces an output $V_o = \frac{(X_1-X_2)(Y_1-Y_2)}{10V}$. If $V_{pulse} = 0$ the circuit is exactly equal to that reported in [23] and the limit cycle shown in Figure 8.5(b) with a frequency $f = \frac{1}{2\pi R_i C_i} \approx 1.6kHz$, with $i = 1,2$ is observed. By varying the potentiometer R_7, a smaller amplitude limit cycle, such as that reported in Figure 8.5(c), is obtained. Moreover, if we now switch on the pulse generator, different dynamics can be obtained. If we introduce a pulse signal of amplitude $1.48V$,

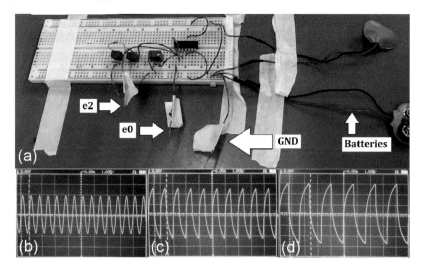

FIGURE 8.3

Hewlett oscillator: (a) physical implementation of the scheme reported in Figure 8.1 by using discrete components, (b) sinusoidal oscillation for low gain, (b) distorted oscillation, and (c) heavily distorted oscillation for higher gains. For oscilloscope traces: horizontal axis $1ms/$div, vertical axis $2V/$div.

duration $200\mu s$ and frequency of $1.13kHz$ the limit cycle reported in Figure 8.5(d) is obtained. If the frequency is $f = 1.07kHz$ the chaotic attractor shown in Figure 8.5(e) can be observed.

This example indicates that the forced van der Pol oscillator is a rich source of different dynamical behaviors.

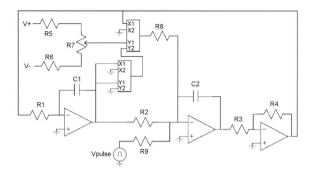

FIGURE 8.4
Implementation of the driven van der Pol oscillator. Component values:
$R_1 = R_2 = R_3 = R_4 = R_9 = 10k\Omega$, $R_5 = R_6 = 470k\Omega$, $R_7 = 100k\Omega$
(potentiometer), $R_8 = 1k\Omega$, $C_1 = C_2 = 10nF$. TL-084 operational amplifiers
and AD633 analog multipliers have been used.

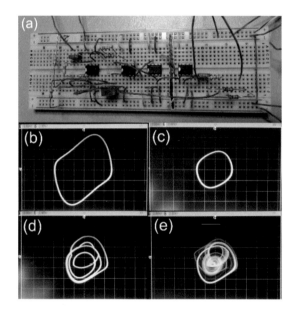

FIGURE 8.5
Van der Pol oscillator with forcing signal: (a) the circuit implemented on
breadboard; (b) limit cycle obtained for $V_{pulse} = 0$ and $R_7 = 100k\Omega$; (c)
limit cycle for $V_{pulse} = 0$ and $R_7 = 50k\Omega$; (d) limit cycle obtained with a
forcing pulse signal with frequency $1.13kHz$; (e) chaotic attractor obtained
with a forcing pulse signal with frequency $1.07kHz$. For oscilloscope traces:
horizontal axis $2V/\text{div}$, vertical axis $2V/\text{div}$.

8.3   An "elegant" oscillator

The oscillator we are reporting here is one of the simplest oscillators that can be conceived with electronic components. Our attention is devoted to understanding if this type of oscillator could become chaotic. The scheme of the circuit is reported in Figure 8.6 and consists mainly of an RC circuit powered by a DC voltage and an NPN 2N222 transistor where only the emitter and the collector are used, leaving the base floating. The genesis of this oscillator is inspired by the Pearson and Anson effect, discovered in 1922. The effect consists in having an oscillation when a powered RC circuit is connected with a neon bulb in parallel to the capacitor. The equivalent effect is obtained by using the tunnel effect in BJT for which Leo Esaki won the Nobel Prize in 1973. The 2N222 transistor is configured to behave as a negative differential resistor (NDR) [80].

An empirical description of the behavior of the RC oscillator is the following. The voltage across the capacitor increases with a time-constant RC until the breakdown voltage of the BJT is reached. At this time the emitter and the collector are short circuited and the capacitor has a fast discharge dynamics and the process cyclically continues. Even if the frequency of the oscillator is mainly linked to the RC time-constant, it also depends on the exact value of the breakdown voltage and, of course, on the discharging time of the capacitor.

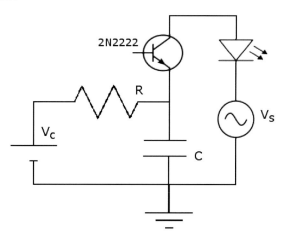

FIGURE 8.6
Implementation of the "elegant" oscillator. Component values: $R = 1.5\mathrm{k}\Omega$, $C = 220\mu F$, 2N222 BJT, $V_c = 10V$.

In the Pearson and Anson oscillator, chaos is observed when a sinusoidal voltage supply is included in series with the neon bulb. The simplicity of the configuration led us, following [80], to include it in the class of "elegant" chaos circuits. Similarly, in our circuit, a sinusoidal voltage supply is put in series to the BJT at the collector. The implementation on breadboard is shown in the picture in Figure 8.7(a). When the external forcing is turned off, i.e., $V_s = 0$, the regular oscillations shown in Figure 8.7(b) are retrieved, while when a sinusoidal voltage $V_s = A \sin(2\pi ft)$ is applied with amplitude $A = 100mV$ and frequency $f = 4$Hz a chaotic oscillation appears, as reported in Figure 8.7(c).

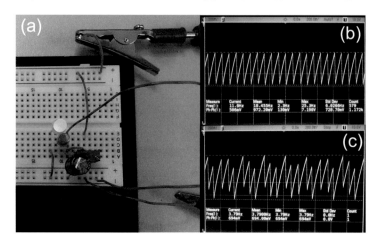

FIGURE 8.7
Elegant oscillator: (a) the circuit implemented on breadboard; (b) temporal trend of the oscillator without external forcing; (c) irregular trend of the oscillations when an external forcing is operating with amplitude $A = 100mV$ and frequency $f = 4$Hz. For oscilloscope traces: horizontal axis $200ms/$div, vertical axis $200mV/$div.

8.4 Synchronization of two Hewlett oscillators

The experiment presented in this section is the first in which a coupling between two circuits is taken into account. Let us consider two Hewlett oscillators implemented according to Section 8.1 and introduce a potentiometer R_c as shown in Figure 8.8.

When the two circuits are coupled through a large resistance, i.e., $R_c = 2$kΩ, they oscillate with the same frequency $f = 1.52$kHz but the two waveforms are out of phase, as reported in Figure 8.9(a). Consequently, the plot $V_{out,1} - V_{out,2}$ shown in Figure 8.9(b) displays a closed curve. Setting the coupling resistor R_c to 150Ω, the two oscillations are synchronized, as shown in Figure 8.9(c)-(d).

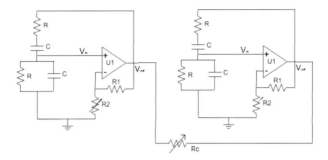

FIGURE 8.8
Two Hewlett oscillators coupled through a diffusion resistor R_c.

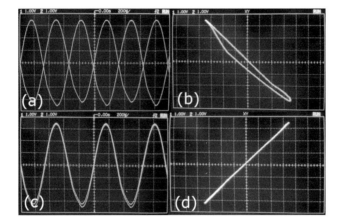

FIGURE 8.9
Behavior of two coupled Hewlett oscillators: temporal trend of $V_{out,1}$ and $V_{out,2}$, and plane $V_{out,1} - V_{out,2}$ when (a)-(b) $R_c = 2k\Omega$ and (c)-(d) $R_c = 150\Omega$. For oscilloscope traces: (a)-(c) horizontal axis $200\mu s/$div, vertical axis $1V/$div; (b)-(d) horizontal axis $1V/$div, vertical axis $1V/$div.

8.5 Multilayer CNN cell with slow-fast dynamics for RD-CNN

In this section, we report the circuit implementation of a two-layer CNN characterized by slow-fast dynamics. The circuit has been realized starting from Equations (6.34) through the schematic reported in Figure 8.10, in which a temporal rescaling $\tau = \kappa t$ with $\kappa = \frac{1}{R_8 C_1} = \frac{1}{R_{21} C_2} \approx 5$ has been introduced. The circuit equations, thus, read as:

$$R_8 C_1 \frac{dx_1}{d\tau} = -x_1 + \frac{R_3}{R_6} y_1 - \frac{R_3}{R_5} y_2 - \frac{R_3}{R_4} \frac{R_2}{R_1 + R_2} V_{cc}$$
$$R_{21} C_2 \frac{dx_2}{d\tau} = -x_2 + \frac{R_{15}}{R_{18}} y_1 + \frac{R_{15}}{R_{19}} y_2 + \frac{R_{15}}{R_{17}} \frac{R_2}{R_1 + R_2} V_{cc} \tag{8.1}$$

while the output of each cell can be written as:

$$y_i = \frac{R_{14}}{R_{13} + R_{14}} \frac{R_9}{R_{11}} V_{sat}(x_i) \tag{8.2}$$

where $V_{sat} = (|x_i + E_{sat}| - |x_i - E_{sat}|)$. Component values have been selected in order to match Equations (8.1) with Equations (6.34), while the component values of the output blocks have been chosen in order to obtain a saturation at $E_{sat} = 1V$. A dual voltage supply is applied as $V_s = \pm 15V$. In order to obtain the slow-fast oscillation, potentiometer R_1 may need a fine tuning around its nominal value.

The circuit implemented on the breadboard shown in Figure 8.11(a) realizes a slow-fast dynamics as shown by the limit cycle reported in Figure 8.11(b) corresponding to the temporal trends of the state variables reported in Figure 8.11(c).

8.6 Non-autonomous multilayer CNN with chaotic behavior

In this section we show the circuit implementation of the two-cells CNN with sinusoidal input introduced in Example 6.13. This circuit is driven by an external input consisting of a sinusoidal signal $V_{in} = A \sin(2\pi f \tau)$, which can be realized by using a function generator or from the line-out of a PC audio board. The frequency of the input signal plays a crucial role in the dynamics shown by the CNN which is able to display limit cycles and, eventually, a chaotic attractor. The scheme reported in Figure 8.12 implements Equations (6.32) introducing a temporal rescaling $\tau = \kappa t$ with $\kappa = \frac{1}{RC} \approx 21000$. The circuit realized on a breadboard is shown in Figure 8.13(a) and its behavior on the $x_1 - x_2$ plane reported in Figure 8.13(b)-(e) for $A = 4V$ and for different values of the frequency f. For $f = 5.3$kHz chaotic motion is observed.

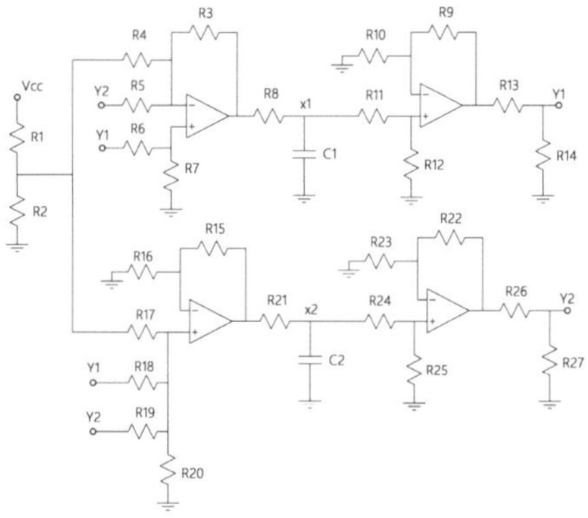

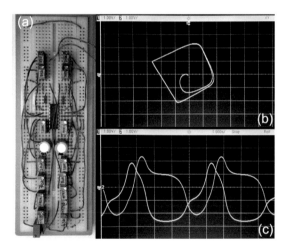

FIGURE 8.10

Circuit scheme for the implementation of the two-layer CNN with slow-fast dynamics. Component values: $R_1 = 13k\Omega$ (potentiometer), $R_2 = 2k\Omega$, $R_3 = R_5 = R_{15} = R_{18} = R_{20} = 120k\Omega$, $R_4 = R_{17} = 820k\Omega$, $R_6 = R_{19} = 7k\Omega$, $R_7 = 270k\Omega$, $R_8 = R_{14} = R_{21} = R_{27} = 1k\Omega$, $R_9 = R_{12} = R_{22} = R_{25} = 1M\Omega$, $R_{10} = R_{11} = R_{23} = R_{24} = 75k\Omega$, $R_{13} = R_{26} = 12.1k\Omega$, $R_{16} = 43k\Omega$, $C_1 = C_2 = 220\mu F$, TL084 op-amps, $V_s = \pm 15V$.

FIGURE 8.11

Two-layer CNN with slow-fast dynamics: (a) circuit implemented using discrete components, (b) time evolution of x_1 and x_2, (c) limit cycle starting from zero initial conditions (capacitors are short-circuited before power supply is turned on). For oscilloscope traces: (b) horizontal axis $1V$/div, vertical axis $1V$/div; (c) horizontal axis $1s$/div, vertical axis $1V$/div.

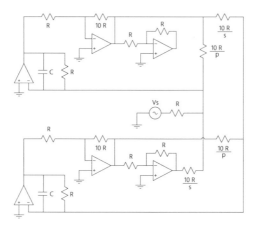

FIGURE 8.12
Circuit scheme for the implementation of the non-autonomous two-layer CNN with chaotic behavior. Component values: $R = 4.75k\Omega$, $s = 1.2$, $p = 1.65$, $C = 10nF$, TL084 OP-AMP, $V_s = \pm 10V$.

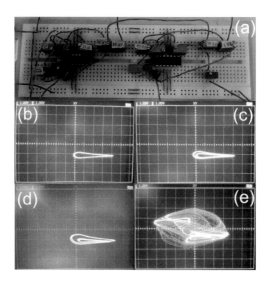

FIGURE 8.13
Non-autonomous two-layer CNN with chaotic behavior: (a) circuit implemented using discrete components, (b) period-1 limit cycle obtained for $f = 4.62$kHz, (c) period-2 limit cycle obtained for $f = 4.83$kHz, (d) period-4 limit cycle obtained for $f = 4.92$kHz, (e) chaotic attractor obtained for $f = 5.3$kHz. For oscilloscope traces: horizontal axis 1V/div, vertical axis 1V/div.

8.7 Multilayer SC-CNN with Chua's circuit dynamics

Consider the electronic scheme reported in Figure 6.9. As has been discussed in Section 6.4.2, it represents the three layers SC-CNN implementing the Chua's circuit dynamics. Note that the nonlinear output stage is realized only for the first cell, implementing the x state variable which actually drives the nonlinearity of the Chua's circuit. Furthermore, the first cell is implemented so that the parameter α can be varied acting on a single potentiometer, namely R_6. A picture of the circuit implemented on a PCB is illustrated in Figure 8.14(a), where the potentiometer is clearly visible. Decreasing R_6, the value of the parameter α is increased, thus following the period-doubling route to chaos passing from a single equilibrium point to limit cycles of different periodicity and eventually single and double-scroll chaotic attractors, as reported in Figure 8.14(b)-(f). A further increasing of the potentiometer leads the circuit towards the outer limit cycle that coexists with the nominal attractor.

8.8 Multilayer SC-CNN implementing hyperchaotic Chua's circuit dynamics

This section is aimed at implementing a hyperchaotic circuit by slightly modifying the SC-CNN implementation of the Chua's circuit dynamics. The equations of the hyperchaotic Chua's circuit read as:

$$\begin{aligned}
\dot{x} &= \alpha(y - h(x)) \\
\dot{y} &= x - y + z + w \\
\dot{z} &= -\beta y + w \\
\dot{w} &= k_1 x + k_2 y + \omega w
\end{aligned} \tag{8.3}$$

with $h(x) = m_1 x + 0.5(m_0 - m_1)(|x+1| - |x-1|)$. Equations (8.3) are derived from those of the original Chua's circuit by adding a further variable w and a feedback term in the second and third equations. Choosing $\alpha = 9.5$, $\beta = 16$, $m_0 = -\frac{1}{7}$, $m_1 = \frac{2}{7}$, $k_1 = -0.1$, $k_2 = 0.6$, and $\omega = 0.03$ the system exhibits two positive Lyapunov exponents.

From an experimental point of view, it is sufficient to include a fourth cell in the original SC-CNN implementation of the Chua's circuit dynamics. The complete circuit is schematized in Figure 8.15, where the fourth cell is designed to match the dynamical equations of the system.

The hyperchaotic attractor of the implemented circuit shown in Figure 8.16(a), is reported on the six possible projections of the state space in Figure 8.16(b)-(g).

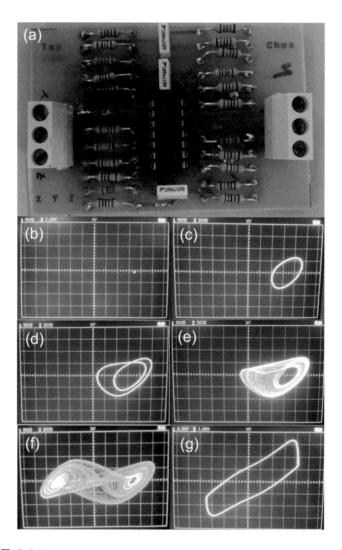

FIGURE 8.14

SC-CNN implementing Chua's circuit dynamics. (a) the circuit implemented
on a PCB, (b) equilibrium point, (c) period-1 limit cycle, (d) period-2 limit
cycle, (e) single-scroll chaotic attractor, (f) double-scroll chaotic attractor,
and (g) limit cycle external to the chaotic attractor. For oscilloscope traces:
(b) horizontal axis $500mV/$div, vertical axis $1V/$div; (c)-(f) horizontal axis
$500mV/$div, vertical axis $200mV/$div; (g) horizontal axis $2V/$div, vertical axis
$1V/$div.

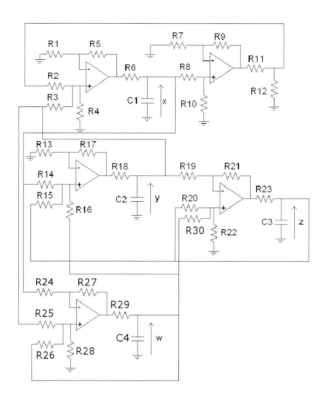

FIGURE 8.15

Circuit scheme for the SC-CNN implementing the hyperchaotic Chua's circuit. $R_1 = 4k\Omega$, $R_2 = 13.3k\Omega$, $R_3 = 5.6k\Omega$, $R_4 = 20k\Omega$, $R_5 = 20k\Omega$, $R_6 = 380\Omega$ (potentiometer), $R_7 = 112k\Omega$, $R_8 = 112k\Omega$, $R_9 = 1M\Omega$, $R_{10} = 1M\Omega$, $R_{11} = 8.1k\Omega$, $R_{12} = 1k\Omega$, $R_{13} = 50k\Omega$, $R_{14} = 100k\Omega$, $R_{15} = 100k\Omega$, $R_{16} = 100k\Omega$, $R_{17} = 100k\Omega$, $R_{18} = 1k\Omega$, $R_{19} = 8.2k\Omega$, $R_{20} = 100k\Omega$, $R_{21} = 100k\Omega$, $R_{22} = 9k\Omega$, $R_{23} = 1k\Omega$, $R_{24} = 100k\Omega$, $R_{25} = 16.7k\Omega$ (potentiometer), $R_{26} = 9.7k\Omega$, $R_{27} = 10k\Omega$, $R_{28} = 18k\Omega$, $R_{29} = 1k\Omega$, $R_{30} = 100k\Omega$, $C_1 = C_2 = C_3 = C_4 = 100nF$. TL084 op-amps, $V_s = \pm 9V$.

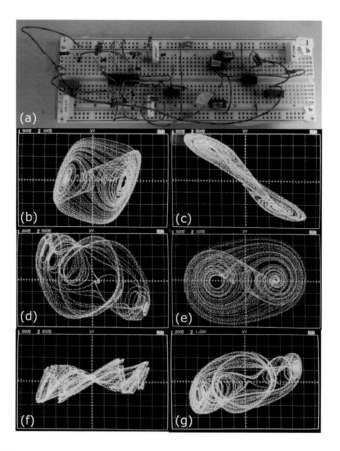

FIGURE 8.16
SC-CNN implementing the hyperchaotic Chua's circuit dynamics. (a) The circuit implemented on the breadboard. Different projections of the hyperchaotic attractor, phase plane: (b) $x - y$ (horizontal axis $500mV/$div, vertical axis $100mV/$div), (c) $x - z$ (horizontal axis $500mV/$div, vertical axis $500mV/$div), (d) $x - w$ (horizontal axis $200mV/$div, vertical axis $500mV/$div), (e) $y - z$ (horizontal axis $500mV/$div, vertical axis $100mV/$div), (f) $y - w$ (horizontal axis $200mV/$div, vertical axis $200mV/$div), (g) $z - w$ (horizontal axis $200mV/$div, vertical axis $1V/$div).

8.9 A SC-CNN chaotic circuit with memristor

This section discusses the SC-CNN implementation of a chaotic circuit in which the nonlinearity is represented by a memristor. Despite the existence of this device being postulated in 1971 by Leon O. Chua [19] as the fourth basic circuit element, memristor characteristics have been effectively found in a real device after 37 years [85]. Nowadays, the memristor has a great theoretical importance as many natural and physical phenomena have memristive behavior. Furthermore, being at the same time a memory and a nonlinear element it is particularly suitable for the implementation of nonlinear, and especially chaotic, circuits. In this section, we present a SC-CNN implementation of a memristive device whose nonlinearity is responsible for the chaotic behavior of a simple nonlinear circuit, namely the memristive Chua's oscillator [10]. The equations of the system read as follows:

$$
\begin{aligned}
\dot{x} &= \alpha(y - x + \xi x - W(w)x) \\
\dot{y} &= x - y + z \\
\dot{z} &= -\beta y - \gamma z \\
\dot{w} &= x
\end{aligned}
\tag{8.4}
$$

where the nonlinearity of the memristor is:

$$
W(w) = \begin{cases} a & \text{if } |w| < 1 \\ b & \text{if } |w| \geq 1 \end{cases}
\tag{8.5}
$$

with $\alpha = 10$, $\beta = 13$, $\gamma = 0.35$, $\xi = 1.5$, $a = 0.3$, and $d = 0.8$.

The implementation of the memristive device, reported in Figure 8.17(a), is based on an open-loop comparator whose output is connected to a high-speed switch acting on the feedback resistor of an operational amplifier, thus varying its gain. The rest of the circuit, shown in Figure 8.17(b), is designed as an SC-CNN whose component values are chosen in order to match circuit equations to Equations (8.4).

The experimental chaotic attractor from the circuit is reported in Figure 8.18 as observed with an oscilloscope on its six projections.

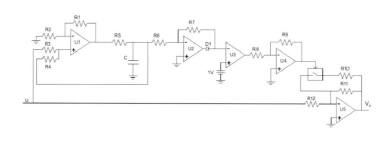

(a)

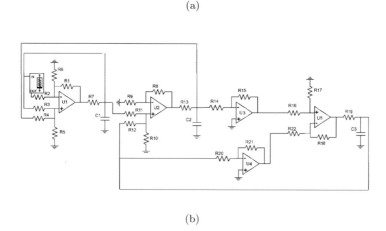

(b)

FIGURE 8.17

(a) Implementation of the memristor, component values: $R_1 = R_3 = R_4 = R_6 = R_7 = 10k\Omega$, $R_2 = 26k\Omega$, $R_5 = R_8 = R_9 = R_{12} = 1k\Omega$, $R_{10} = 480\Omega$, $R_{11} = 800\Omega$, $C = 100nF$, $D1$ is a $2N222$ diode, op-amps are $TL084$, the switch is a $ADG201AKN$. (b) Implementation of the memristive Chua's oscillator, component values: $R_1 = R_8 = R_{11} = R_{15} = R_{16} = R_{17} = R_{18} = R_{21} = R_{22} = 100k\Omega$, $R_2 = R_4 = R_6 = 10k\Omega$, $R_3 = 16.6k\Omega$, $R_5 = 20k\Omega$, $R_7 = R_{13} = R_{19} = 1k\Omega$, $R_9 = 25k\Omega$, $R_{10} = R_{12}50k\Omega$, $R_{14} = 15.38k\Omega$, $R_{20} = 153.8k\Omega$, $C1 = C2 = C3 = 100nF$, op-amps are $TL084$.

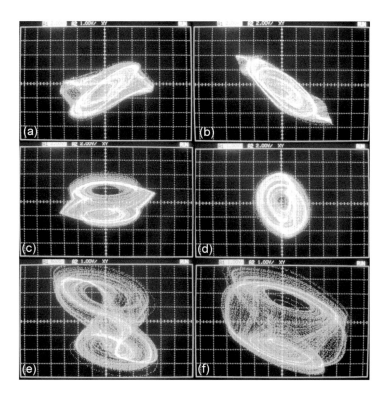

FIGURE 8.18

SC-CNN implementing the memristive Chua's oscillator dynamics. Projection of the attractor on the phase planes: (a) $x - y$ (horizontal axis $2V/$div, vertical axis $1V/$div); (b) $x - z$ (horizontal axis $2V/$div, vertical axis $2V/$div); (c) $x - w$ (horizontal axis $2V/$div, vertical axis $2V/$div); (d) $y - z$ (horizontal axis $1V/$div, vertical axis $2V/$div); (e) $y - w$ (horizontal axis $500mV/$div, vertical axis $1V/$div); (f) $z - w$ (horizontal axis $1V/$div, vertical axis $1V/$div).

8.10   Synchronization of two Chua's dynamics with diffusive coupling

Let us consider two diffusively coupled SC-CNNs implementing the Chua's circuit as in Figure 6.18. The coupling is regulated by the value of resistor R_c such that if $R_c \to \infty$ the two circuits are uncoupled and their behavior is unsynchronized. If $R_c \to 0$ a strong coupling is realized and synchronization appears.

In the experiment assume first $R_c = 5\text{k}\Omega$. In this case the two Chua's dynamics are weakly coupled and still not synchronized. Starting now to decrease the coupling resistor in order to reach synchronization, a range of values can be found which drive the two circuits towards a limit cycle. This occurs around $R_c = 1\text{k}\Omega$ and it is due to the specific physical implementation which admits the coexistence of a limit cycle external to the nominal attractor. In order to escape from such a regime, the capacitors are short-circuited so as to reset their initial conditions. Further decreasing of the coupling resistor leads to the synchronous regime reported in Figure 8.19 which has been obtained for $R_c = 330\Omega$.

The analytical computation of the synchronization threshold can be obtained by the MSF approach [9], as discussed in Section 7.4.

It is interesting to note that in this experiment a minimum in the power absorbed by the two SC-CNNs is reached at the onset of the synchronous oscillations.

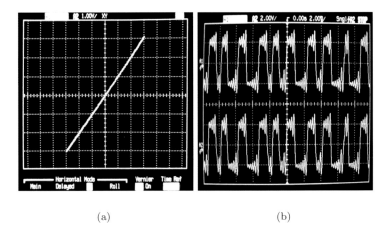

(a)   (b)

FIGURE 8.19
Experimental synchronization of two SC-CNN based Chua's circuits: (a) phase plane y_1 vs. y_2 (horizontal axis $1V/\text{div}$, vertical axis $1V/\text{div}$), (b) waveforms related to state variables z_1 and z_2 (horizontal axis $2ms/\text{div}$, vertical axis $2V/\text{div}$).

8.11   Chaos encryption

The experiment discussed in this section is related to the implementation of a secure communication system based on chaotic encryption and synchronization. We used as chaos generator the Chua's circuit implemented with a three-layer SC-CNN, as discussed in Section 6.4.2. The system is mainly composed of two units: a transmitter, in which the information is opportunely masked, and a receiver which is able to synchronize with the transmitter and to decode the hidden information. The schematic representation reported in Figure 8.20 shows the complete equipment of the transmitter station, which includes a Chua's circuit and a modulator.

The modulator, whose electrical scheme is reported in Figure 8.21, acts on the input signal, which is a voltage suitably amplified, providing a current proportional to it which is then injected on the capacitor storing the first state variable of a Chua's circuit, i.e., C_1 in the scheme of Figure 6.9. The voltage across C_1 is then sent through the transmission line to the receiver.

The receiver is schematized in Figure 8.22 in which a second Chua's circuit with the same parameters as the first one is hosted. Despite being identical, the two Chua's circuits are unsynchronized, since they start from different initial conditions. However, synchronization can be attained applying a master-slave scheme based on the inverse system as discussed in Section 7.3.3. The receiver is also equipped with a demodulator, whose electrical scheme is reported in Figure 8.23, which takes as input the voltage across a shunt resistor, representing the difference between the two state variables. The error signal, once synchronization is attained, is given only by the current injected in the transmitter containing the hidden information, therefore the demodulator transforms the differential voltage across the shunt resistance set to 10Ω to a single-ended one. A final filtering stage, using a simple low-pass RC circuit, is used to attenuate the high-frequency noise in the decoded information.

In our experiment, we used an audio signal as given by a standard PC audio board and the output of the filter stage in the receiver is provided to a loudspeaker. A picture of the experimental setup is reported in Figure 8.24. In the circuits TL084 operational amplifiers have been used.

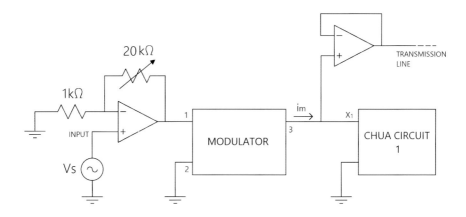

FIGURE 8.20
Schematic representation of the transmitter.

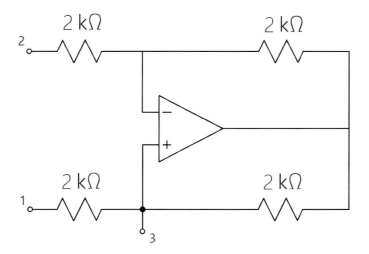

FIGURE 8.21
Electrical scheme of the modulator.

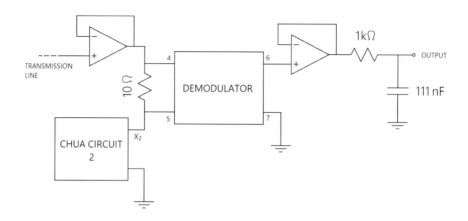

FIGURE 8.22
Schematic representation of the receiver.

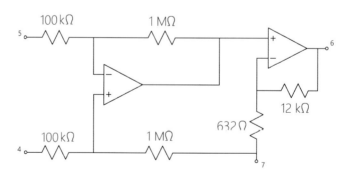

FIGURE 8.23
Electrical scheme of the demodulator.

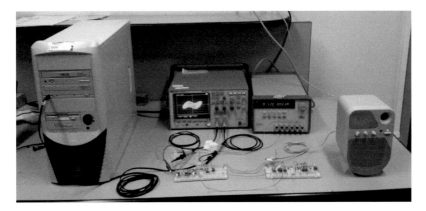

FIGURE 8.24
Experimental setup for the chaos encryption experiment: the PC audio board is used to generate the information which is passed to the circuitry, the loudspeaker reproduces the information as reconstructed by the receiver.

8.12 Qualitative chaos-based sensors

The experiment reported in this section aims to show how a chaotic circuit can be used to obtain a qualitative sensor. Let us start from the SC-CNN realization of the Chua's circuit dynamics, reported in Figure 6.9. As discussed in the experiment in Section 8.7, the potentiometer R_6 is used to implement the parameter α of the Chua's circuit which drives the dynamical behavior of the system through a series of bifurcations leading to chaos. We now replace the potentiometer R_6 with the device shown in Figure 8.25(a). It represents a two-terminal element constituted by an electrolytic cell containing water and two copper electrodes located in parallel and immersed in the liquid. The dimensions of each electrode are $W = 20$mm and $L = 50$mm, and the distance between the two electrodes is $d = 35$mm. The maximum level of liquid in the cell is 30mm.

The liquid level modifies the equivalent resistance of the electrolytic cell and, hence, can be used to vary the value of parameter α. An empty cell corresponds to an infinite resistance whose value is decreased by increasing the liquid level. The whole bifurcation scenario described in Section 8.7 can be observed by slowly filling the cell. The experimental setup is reported in Figure 8.25(a), while two examples of a chaotic attractor obtained at different levels of the solution in the cell are reported in Figure 8.25(b)-(c).

It is interesting to note that this experiment realizes a chaos-based qualitative sensor of the water: on the basis of the electrochemical properties of the water, in fact, the same quantity of liquid in the cell generates different values of the equivalent resistance. Thus, by looking at the dynamical behavior shown by the chaotic circuit, it is possible to qualitatively discern different waters. Let us consider waters collected from the public water supply system in different towns of eastern Sicily, as depicted in Figure 8.26. We measure the level of water needed to obtain a double-scroll attractor when filling the cell with the different waters. The collected measures are reported in Table 8.1. The wide range of values obtained is due to the highly different electrochemical properties of the sources of water: the lower is the conductivity of the water; the higher is the level needed to obtain the same equivalent resistance.

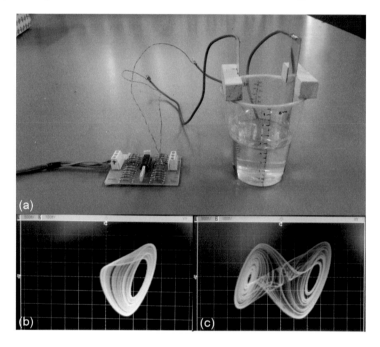

FIGURE 8.25
Chua's circuit with electrolytic cell: (a) circuit implemented with the water cell, (b) single-scroll attractor and (c) double-scroll attractor obtained with different levels of water in the cell. For oscilloscope traces: horizontal axis $500mV/\text{div}$, vertical axis $100mV/\text{div}$.

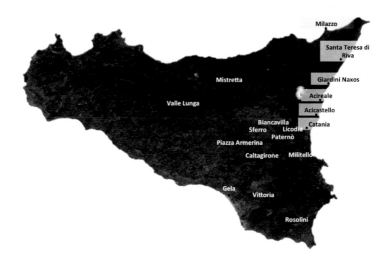

FIGURE 8.26
Dislocation of the towns considered in the experimental evaluation of water electrochemical properties in eastern Sicily.

TABLE 8.1
Water Level (in mm) in the Cell to Obtain a Double-Scroll Attractor in the Chua's Circuit for Different Towns in Sicily

Town	water level (mm)	Town	water level (mm)
Acicastello	84	Militello Val di Catania	103
Acireale	100	Mistretta	> 260
Biancavilla	55	Paternó	61
Caltagirone	127	Piazza Armerina	76
Catania (perifery)	60	Rosolini	175
Catania (downtown)	63	Santa Maria di Licodia	64
Catania (industrial area)	52	Santa Teresa di Riva	≫ 260
Gela	78	Sferro	15
Giardini Naxos	68	Vallelunga Pratameno	> 260
Milazzo	> 260	Vittoria	62

8.13   Chaos control experiment

In this section an experimental strategy to control chaos is presented. It is based on the use of an impulsive signal transmitted to the controlled circuit by means of a transformer. This control strategy is based on the introduction of a small perturbation to the bifurcation parameter, resembling classical chaos control techniques which are based on parameter perturbation with sinusoidal signal, but in this case the control is performed acting directly at circuit level. Moreover, the use of a transformer allows us to decouple the main circuit from the control signal generator.

Consider the SC-CNN implementing the Chua's circuit dynamics (Figure 6.9) and focus on the cell implementing the first state variable. The transformer, whose voltage ratio is $\frac{1}{10}$, is introduced according to the scheme reported in Figure 8.27, thus the secondary winding is put in series with the RC group at the output of the operational amplifier. The primary winding of the transformer is connected to a signal generator providing a square pulse wave, with an amplitude $A = 1V$ and a 20% duty cycle. Varying the frequency of this control signal, the different limit cycles included in the chaotic attractor can be stabilized, as the results presented in Figure 8.28 clearly demonstrate.

Opportunely setting the bifurcation parameter by varying R_6 it is possible to drive the circuit towards limit cycles of different periodicity. We consider these limit cycles as different *genes* on which the control law is applied. In Figure 8.28 each row reports the effect of the control law on the given gene for different values of the frequency of the square pulse wave connected to the primary winding of the transformer. A wide range of limit cycles can be effectively stabilized by means of this simple and non-invasive control strategy, which can be easily applied to other chaotic circuits implemented following the SC-CNN approach.

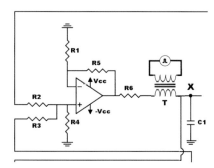

FIGURE 8.27
Circuit scheme for the control strategy obtained including transformer T in the first CNN cell of the Chua's circuit implementation reported in Figure 6.9.

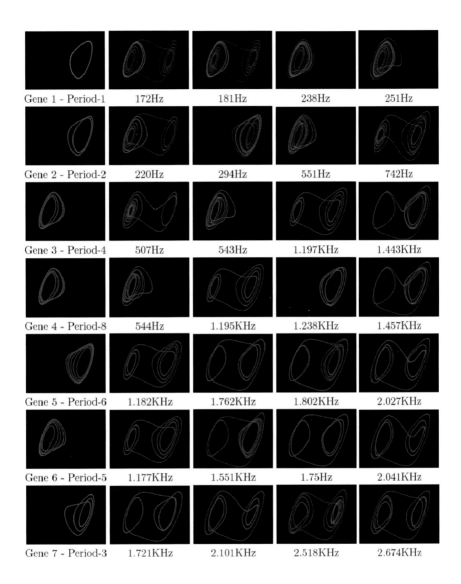

FIGURE 8.28

Controlling chaos in Chua's circuit. Different limit cycles obtained starting from seven genes, corresponding to a particular dynamical behavior of the Chua's circuit and selected by varying the bifurcation parameter, at the given frequency of the control signal.

8.14   Logistic map with Arduino

In this section we discuss an experiment aimed at realizing a discrete-time chaotic system, i.e., the logistic map introduced in Chapter 2. Discrete-time systems are easily implemented in digital computing devices, with the only constraint being the used digit precision. The MATLAB® procedure calculating the logistic map is itself a way to implement it on a standard PC. However, we want to implement a circuit which produces an output voltage which follows the dynamics of a logistic map.

The circuit discussed in this section is based on a well-known microcontroller board: the Arduino® UNO. It is a multipurpose board mounting an ATMEL 3-2bit ARM-based processor interfaced with several components, including an analog-digital converter and serial/USB interface. It is fully programmable via an integrated development environment in C/C++. Arduino® boards are often distributed in educational kits which include the board, a series of standard components, a breadboard, and a self-contained operation manual in which a series of comprehensive sample projects are described. The experiment shown in this section is therefore discussed following the structure of the Arduino® operation manual.

Ingredients:

- seven $10k\Omega$ resistors;

- nine $20k\Omega$ resistors;

- a TL084 operational amplifier;

- a dual-voltage power supply;

- wires.

Time: 45 minutes.

Experiment description: the microcontroller is programmed so that the logistic iterator is implemented and the result coded as an 8-bit word and sent through the digital output of the board. These digital signals are then converted to an analog voltage by means of a standard $R - 2R$ converter; the output of the converter is fed back as input of the microcontroller analog-digital converter and used to further iterate the logistic map. The scheme is reported in Figure 8.29.

Building the circuit: let us start from the Arduino® board. We use pin $A3$ as input, while pins from 2 to 9 are the output eight digits of the 8-bit word. The $R - 2R$ digital-to-analog converter is realized on a breadboard following the schematic in Figure 8.30. Wires connect the output pins of the Arduino® board with the inputs of the converter in such a way that the most significant bit (corresponding to pin 2) is connected to resistor R_1, and so on. Finally,

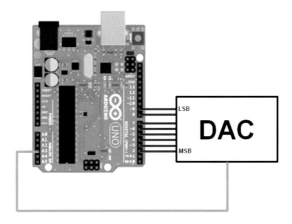

FIGURE 8.29

Schematic representation of the experimental setup: the Arduino® board reads an analog input constituting the analog representation of the digital output of the board.

the output of the converter is connected to the input pin $A3$. The complete experimental setup is shown in Figure 8.31.

The code: the microcontroller is programmed with the following procedures:

```
int inPin = A3; //analog input pin 3
int outPin1 = 2; //digital output pin 2
int outPin2 = 3; //digital output pin 3
int outPin3 = 4; //digital output pin 4
int outPin4 = 5; //digital output pin 5
int outPin5 = 6; //digital output pin 6
int outPin6 = 7; //digital output pin 7
int outPin7 = 8; //digital output pin 8
int outPin8 = 9; //digital output pin 9
int val = 10; //variable to store the read value
int val1 = 120; //variable to store the value to be written
float xi = 0; //variable converting the read value
float x=0.1; //variable to store iterator result

void setup()
{
// sets the pin as output
pinMode(outPin1, OUTPUT);
pinMode(outPin2, OUTPUT);
pinMode(outPin3, OUTPUT);
pinMode(outPin4, OUTPUT);
pinMode(outPin5, OUTPUT);
pinMode(outPin6, OUTPUT);
pinMode(outPin7, OUTPUT);
pinMode(outPin8, OUTPUT);
// initialize the first output
for (int i=0;i<8;i++) {
  digitalWrite(i+2, val1%2);
  val1=val1/2;
}
}
```

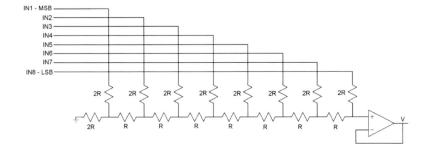

FIGURE 8.30
Digital-to-analog converter based on the $R - 2R$ configuration with output buffer: $R = 10k\Omega$, TL-084 operational amplifiers powered with a dual-voltage supply $V_s = \pm 9V$.

FIGURE 8.31
Complete setup for the logistic map microcontroller implementation with the Arduino® board and the $R - 2R$ converter implemented on the breadboard.

```
void loop() //main function
{
val = analogRead(inPin); // read from the input pin A3
xi =(val) / 1023.0; // normalize in the range [0;1]
x=4.0*xi*(1.0-xi); // iterate
val = 255 * (x); // normalize in the range [0;255]
// write the output
val1=val;
for (int i=0;i<8;i++) {
  digitalWrite(i+2, val1%2);
  val1=val1/2;
}
}
```

The above C++ procedure is composed of three parts. The first part defines and initializes the variables, including those used to address input/output pins. The second part sets up the board enabling the 8 output pins and writing the first output, i.e., the initial condition of the logistic map. This latter value can be modified in the first part of the procedure. The third part is the main loop repeated with a default frequency of about 420kHz. The clock frequency can be suitably modified with further built-in commands. Readers can find these commands in the advanced operation manual available at [2]. At each iteration, the input voltage, the result of the digital-to-analog conversion, is acquired through the analog-to-digital converter and stored in an integer number in the range $[0; 1023]$. This value is normalized in the range $[0; 1]$ and applied to the logistic iterator. The calculated value is converted in an integer number in the range $[0; 255]$ whose 8-bit digital representation is written on the eight output pins.

Experimental results: the setup described above is able to generate the chaotic signal shown in Figure 8.32. Looking at the oscilloscope trace in Figure 8.32(a), it is possible to see that each value in the range $[0V; 5V]$ is maintained for a clock period and then modified according to the result of the logistic iterator. The erratic behavior of the logistic map is clearly seen from a longer acquisition provided at a sampling rate equal to the clock frequency, as shown in Figure 8.32(b). Despite the fact that a precision limited to 8-bit has been used, the chaotic behavior of the logistic map can be observed. It is important to stress that this experiment relies on simple components and on a low cost and easy-to-use microcontroller board, which is becoming widely adopted in home laboratories.

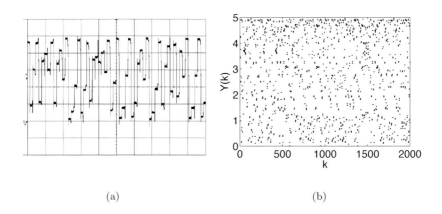

(a)   (b)

FIGURE 8.32
Experimental trend of the logistic map implemented on the Arduino® board: (a) oscilloscope trace ($1V/div$), (b) acquired waveform.

8.15   Networks of SC-CNN Chua's circuits

In this section, a modular network of SC-CNN Chua's circuits is discussed. As schematized in Figure 8.33, the network composed of 9 Chua's circuits coupled by means of diffusion implemented through coupling resistors, similar to what was discussed in Section 8.10 where only two circuits were considered. In this case, if a lattice structure is considered, we have 12 links, each associated to a coupling resistor implemented through a $1k\Omega$ potentiometer. Furthermore, on each Chua's circuit a further potentiometer allows us to set the system parameter α, so that each circuit can exhibit different attractors.

We remark that for all the experiments discussed in the following, when coupling the circuits, it may occur that some of them undergo the periodic regime of the limit cycle external to the chaotic attractor. However, since the value of 150Ω for the coupling resistors is suitable to obtain chaos synchronization in all the considered configurations, it is sufficient to reset capacitors in order to reach the chaotic attractor.

We start considering a network of 9 circuits in double-scroll, as reported in Figure 8.34. When the circuits are coupled in the lattice structure through resistors all set to 150Ω, a synchronized behavior emerges as shown in Figure 8.35, where in the oscilloscope the plots $x_i - x_1$ are reported for $i = 1, \ldots, 9$ numbering circuits in rows starting from the upper-left corner. The behavior of each circuit is still in double-scroll as shown in Figure 8.35.

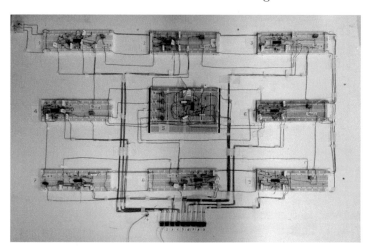

FIGURE 8.33

Experimental setup of 9 Chua's circuits connected in a lattice structure. Each connection can be enabled/disabled and the corresponding coupling coefficient can be tuned acting on a single potentiometer.

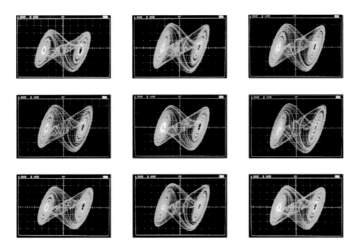

FIGURE 8.34
Attractor of the 9 Chua's circuits when uncoupled. Scales: horizontal axis
$500mV$/div, vertical axis $100mV$/div.

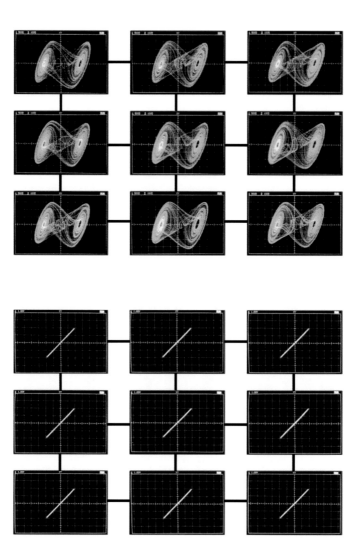

FIGURE 8.35
Attractor of the 9 Chua's circuits when coupled into a lattice structure
with diffusion resistor set at $R = 150\Omega$ (upper panel, scales: horizontal axis
$500mV/$div, vertical axis $100mV/$div), phase planes $x_1 - x_i$ for $i = 1, \ldots, 9$
(lower panel, scales: horizontal axis $1V/$div, vertical axis $1V/$div).

8.15.1 Experiment series 1 — parametric uncertainties

In order to test the robustness of the network to parametric differences, we change the dynamical behavior of some circuits in the network by varying the value of α in given circuits. The coupling resistors are always all set to 150Ω.

8.15.1.1 Configuration 1

Let us introduce four circuits in single-scroll according to the configuration reported in Figure 8.36. When circuits are coupled, a synchronized behavior emerges as shown in Figure 8.37. Interestingly, each circuit exhibits a double-scroll as shown in Figure 8.37. Hence, the four circuits subjected to a parametric uncertainty are forced to follow the behavior of the other five circuits.

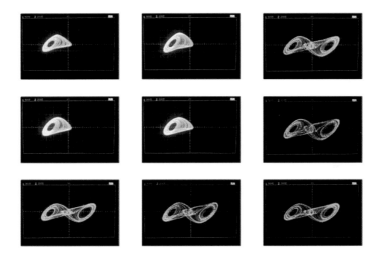

FIGURE 8.36
Attractor of the 9 Chua's circuits when uncoupled: four in single-scroll and five in double-scroll. Scales: horizontal axis $500mV/$div, vertical axis $200mV/$div.

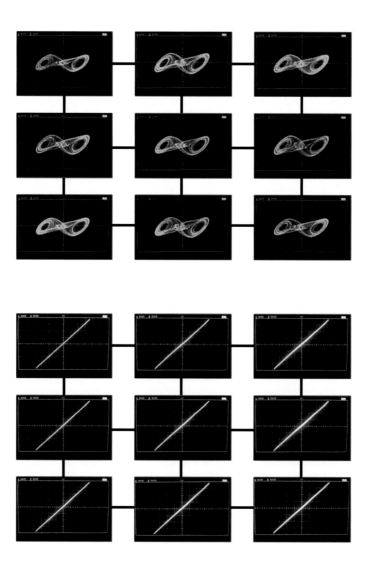

FIGURE 8.37
Attractor of the 9 Chua's circuits when coupled into a lattice structure
with diffusion resistor set at $R = 150\Omega$ (upper panel, scales: horizontal axis
$500mV/\text{div}$, vertical axis $200mV/\text{div}$), phase planes $x_1 - x_i$ for $i = 1, \ldots, 9$
(lower panel, scales: horizontal axis $500mV/\text{div}$, vertical axis $500mV/\text{div}$).
The circuits start from the configuration of Figure 8.36.

8.15.1.2 Configuration 2

Consider now five circuits in single-scroll placed in the network as reported in Figure 8.38. In this case, while the network still ensures a synchronized behavior the whole network oscillates in the regime of single-scroll as shown in Figure 8.39.

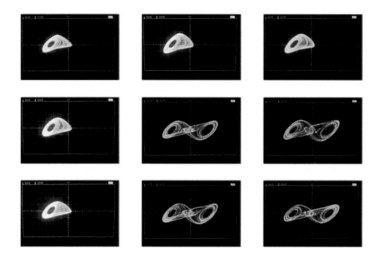

FIGURE 8.38

Attractor of the 9 Chua's circuits when uncoupled: five in single-scroll and four in double-scroll. Scales: horizontal axis $500mV/\text{div}$, vertical axis $200mV/\text{div}$.

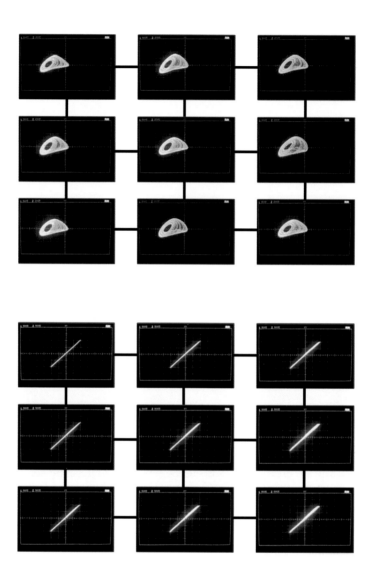

FIGURE 8.39
Attractor of the 9 Chua's circuits when coupled into a lattice structure with diffusion resistor set at $R = 150\Omega$ (upper panel, scales: horizontal axis $500mV/$div, vertical axis $200mV/$div), phase planes $x_1 - x_i$ for $i = 1, \ldots, 9$ (lower panel, scales: horizontal axis $500mV/$div, vertical axis $500mV/$div). The circuits start from the configuration of Figure 8.38.

8.15.1.3 Configuration 3

In this further experiment the effect of the parametric uncertainty is increased, leading to a configuration, reported in Figure 8.40, in which five circuits exhibit a period-1 limit cycle. When coupled in the network all circuits are driven towards a period-1 limit cycle, as shown in Figure 8.41; however, synchronization appears to be affected by the introduced uncertainty. The straight line characterizing the complete synchronization regime now becomes a closed curve denoting a weaker form of synchronization, i.e., lag synchronization, as reported in Figure 8.41.

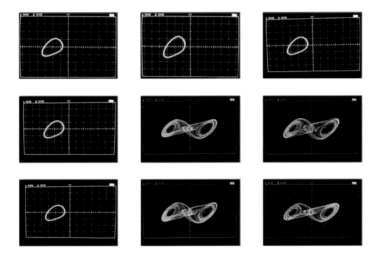

FIGURE 8.40
Attractor of the 9 Chua's circuits when uncoupled: five in period-1 limit cycles and four in double-scroll. Scales: horizontal axis $500mV/$div, vertical axis $200mV/$div.

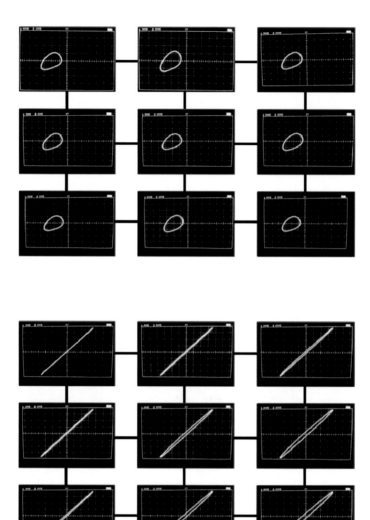

FIGURE 8.41
Attractor of the 9 Chua's circuits when coupled into a lattice structure
with diffusion resistor set at $R = 150\Omega$ (upper panel, scales: horizontal axis
$500mV/\text{div}$, vertical axis $200mV/\text{div}$), phase planes $x_1 - x_i$ for $i = 1, \ldots, 9$
(lower panel, scales: horizontal axis $200mV/\text{div}$, vertical axis $200mV/\text{div}$.).
The circuits start from the configuration of Figure 8.40.

8.15.1.4 Configuration 4

Finally, we focus on the case in which only three circuits are subjected to a heavy uncertainty and driven to an equilibrium point. Consider the configuration shown in Figure 8.42 and couple the circuits according to the considered network structure. Circuits now exhibit a period-2 limit cycle and they are phase synchronized according to the phase plane as shown in Figure 8.43.

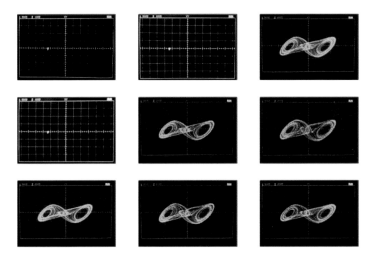

FIGURE 8.42
Attractor of the 9 Chua's circuits when uncoupled: three in equilibrium point and six in double-scroll. Scales: horizontal axis $500mV/\text{div}$, vertical axis $200mV/\text{div}$.

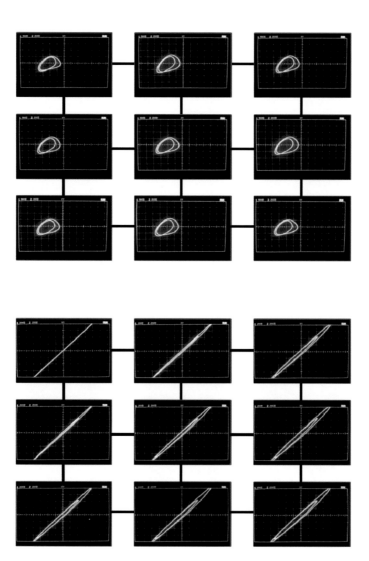

FIGURE 8.43

Attractor of the 9 Chua's circuits when coupled into a lattice structure with diffusion resistor set at $R = 150\Omega$ (upper panel, scales: horizontal axis $500mV/\text{div}$, vertical axis $200mV/\text{div}$), phase planes $x_1 - x_i$ for $i = 1, \ldots, 9$ (lower panel, scales: horizontal axis $200mV/\text{div}$, vertical axis $200mV/\text{div}$). The circuits start from the configuration of Figure 8.42.

8.15.2   Experiment series 2 — different topologies

We focus now on a complementary scenario, namely on the case in which all circuits are in double-scroll, but some of the 12 links are removed. When the link is active the corresponding coupling resistor is set again to 150Ω.

8.15.2.1   Configuration 1

In the first experiment two links are removed, as shown in Figure 8.44. The global behavior of the network is non trivial as reported in Figure 8.44. Synchronization is sensibly affected; even physically coupled circuits display a degraded synchronization as it clearly appears from the thickness of the trajectory on the phase planes $x_1 - x_i$ for $i = 1, \ldots, 9$. Interestingly, circuit 1 and circuit 7 are synchronized with each other but not with circuit 4 through which they are coupled. Besides circuits 1 and 4, circuits 3 and 9 are synchronized, but not with the rest of the network. Finally, circuits 2, 5, and 7 show synchronization but remain weakly synchronized with the other circuits in the network. Hence, this scenario leads to the coexistence of different levels of synchronization in the structure which appears to be not driven by the physical links.

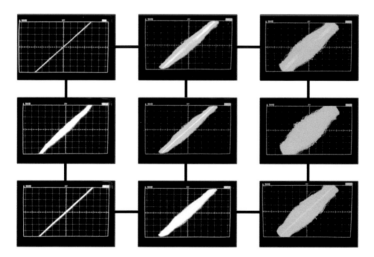

FIGURE 8.44
Synchronization of the 9 Chua's circuits when coupled through the given network: phase planes $x_1 - x_i$ for $i = 1, \ldots, 9$. Scales: horizontal axis $500mV/$div, vertical axis $500mV/$div.

8.15.2.2   Configuration 2

In the second experiment an array configuration, such as that shown in Figure 8.45, is considered. Synchronization is clearly visible, even if the first and the last circuit in the array (upper left corner and lower right corner) appear less synchronized, as indicated by the thickness of the trajectory on the phase planes $x_1 - x_i$ for $i = 1, \ldots, 9$.

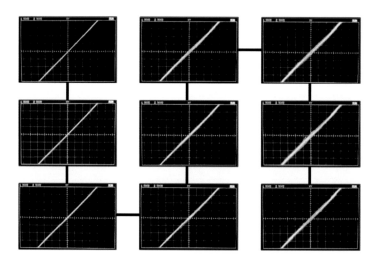

FIGURE 8.45

Synchronization of the 9 Chua's circuits when coupled in array: phase planes $x_1 - x_i$ for $i = 1, \ldots, 9$. Scales: horizontal axis $500mV$/div, vertical axis $500mV$/div.

8.15.2.3   Configuration 3

Let us now consider a closed array, i.e., a ring of Chua's circuits, as reported in Figure 8.46. Closing the ring has a positive effect on the global synchronization which is now comparable with that observed in the original lattice structure which, however, has more links.

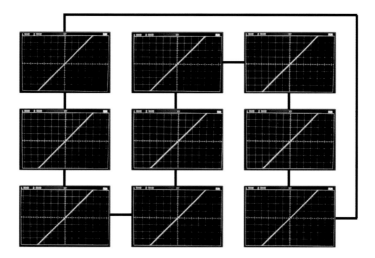

FIGURE 8.46
Synchronization of the 9 Chua's circuits when coupled in a ring: phase planes $x_1 - x_i$ for $i = 1, \ldots, 9$. Scales: horizontal axis $500mV/\text{div}$, vertical axis $500mV/\text{div}$.

8.15.2.4   Configuration 4

As a last example, consider now a star configuration, i.e., a configuration in which the central circuit is linked to all the other Chua's circuits, whose results are reported in Figure 8.47. Despite few links being considered in this scenario, a global synchronization is still attained.

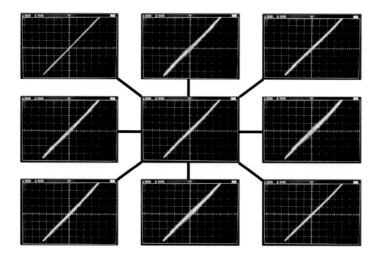

FIGURE 8.47

Synchronization of the 9 Chua's circuits when coupled in a star network: phase planes $x_1 - x_i$ for $i = 1, \ldots, 9$. Scales: horizontal axis $500mV/\mathrm{div}$, vertical axis $500mV/\mathrm{div}$.

8.15.3   Experiment series 3 — effect of an external noise

This last series of experiments performed on the 9 Chua's circuits is aimed at investigating the role that noise can play in networks of nonlinear circuits [11, 13]. Noise in nonlinear complex systems in fact may induce non-trivial behavior. In some cases the injection of an external noise may be beneficial. An example is stochastic resonance where intentionally adding noise to a system produces the amplification of an otherwise weak signal [34].

In the following experiments, we will show the effect of a Gaussian noise injected into one node of the Chua's circuits network on the emerging global behavior and, especially, on synchronization, showing that under some conditions noise leads to significant changes in the whole network dynamics. The noise is realized by a function generator; it is a zero mean Gaussian noise whose variance can be opportunely tuned. We inject noise in the operational amplifier realizing the first state equation of a given node by adding a $20k\Omega$ resistance on the inverting terminal. In order to respect the gain rule a further $20k\Omega$ resistance referred to ground is added on the non-inverting terminal.

8.15.3.1   Configuration 1

In the first configuration we consider the lattice structure of 9 Chua's circuits showing a double-scroll chaotic attractor as reported in Figure 8.34. Let us now inject the noise in one of the corner nodes, namely node 7. Increasing the variance of the noise up to $9.3V$, that is including a noise comparable to the original dynamics, the whole network maintains the synchronous behavior, as shown in Figure 8.48(a) where the plane x_1 vs. x_7. Furthermore, the chaotic attractor shown by the circuit in node 7 reported in Figure 8.48(b) is still a double-scroll despite the presence of a visible noise effect. It is worth noting that if we isolate node 7 from the rest of the network, the injection of the noise with a variance of $9.3V$ drives the circuit out of the chaotic attractor toward the external limit cycle, as reported in Figure 8.49. Thus, in this case, the network has a stabilizing effect, reducing the impact of noise injected through circuit 7.

8.15.3.2   Configuration 2

The second experiment starts from the 9 Chua's circuits that, uncoupled, display unsynchronized single-scroll attractors, while when coupled in the lattice configuration synchronize. In this case, the effect of a noise injected in node 7 is peculiar. The synchronization of the 9 single-scroll attractors is maintained increasing the variance of the noise up to $8V$. After this threshold, the effect of the noise is to drive the whole network towards synchronized double-scroll chaotic attractors, as can be noticed in Figure 8.50.

Interestingly, the threshold value of the variance after which a transition of the whole network from synchronized single-scroll to synchronized double-scroll depends on the number of circuits showing a single-scroll attractor in the

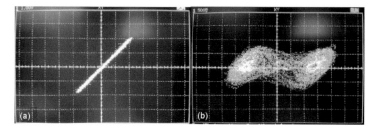

FIGURE 8.48

Synchronization of the 9 Chua's circuits with injection of a Gaussian noise with variance $9.3V$ on node 7: (a) phase plane $x_1 - x_7$ (horizontal axis $1V/$div, vertical axis $1V/$div), (b) chaotic attractor on the plane $x_7 - y_7$ (horizontal axis $500mV/$div, vertical axis $200mV/$div).

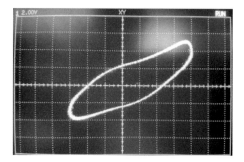

FIGURE 8.49

Effect of a noise with variance $9.3V$ on node 7 when uncoupled from the rest of the network: limit cycle on the plane $x_7 - y_7$. Scales: horizontal axis $2V/$div, vertical axis $1V/$div.

uncoupled configuration. If we set the potentiometer R_6 only in two circuits, namely circuit 5 and circuit 6, so that they show the double-scroll, the lattice configuration without noise drives the circuits to a synchronous single-scroll attractor. However, the injection of a noise with variance $800mV$ in node 7 is now sufficient to move the whole network to the double-scroll as shown in Figure 8.51.

Readers are invited to verify this property by varying the number of circuits showing a double-scroll in the uncoupled configuration and to calculate the critical value of the noise variance.

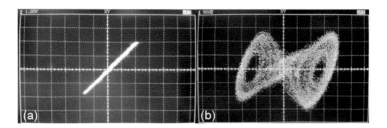

FIGURE 8.50
Synchronization of the 9 Chua's circuits with injection of a Gaussian noise
with variance $8V$ on node 7, when all circuits are identical and synchronized
in single-scroll: (a) phase plane $x_1 - x_7$ (horizontal axis $1V/$div, vertical axis
$1V/$div), (b) chaotic attractor on the plane $x_7 - y_7$ (horizontal axis $500mV/$div,
vertical axis $100mV/$div).

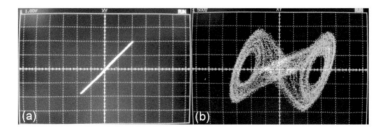

FIGURE 8.51
Synchronization of the 9 Chua's circuits with injection of a Gaussian noise
with variance $800mV$ on node 7, when circuits 5 and 6 are in double-scroll
if uncoupled: (a) phase plane $x_1 - x_7$ (horizontal axis $1V/$div, vertical axis
$1V/$div), (b) chaotic attractor on the plane $x_7 - y_7$ (horizontal axis $500mV/$div,
vertical axis $100mV/$div).

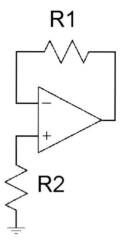

FIGURE 8.52
Circuit for Exercise 8.8.

8.16   Exercises

1. Realize the electronic circuit discussed in Exercise 9 of Chapter 4.

2. Consider the "elegant" oscillator of Section 8.3 and estimate the frequency by changing the 2N222 transistor. Repeat the experiment several times and study the statistics of the value obtained.

3. Realize the chaotic Colpitts oscillator by using the CNN approach.

4. Consider the forced 2-cell CNN with chaotic behavior and synchronize a pair of these circuits.

5. Build two Lorenz oscillators by using electronic components and synchronize them.

6. Repeat the synchronization scheme proposed by Pecora and Carroll with two Rössler electronic circuits.

7. Apply to the Lorenz system the idea of the chaos control scheme to stabilize periodic orbits in the Chua's circuit discussed in Section 8.13.

8. Consider the circuit in Figure 8.52. Verify that chaos can be generated by using an operational amplifier of type $LF357N$. Then, connect several of them in a ring configuration and synchronize the network.

9. Build a network of Lorenz oscillators and perform experiments similar to those reported in Section 8.15 for the Chua's oscillators.

10. Study the effects of noise in the network realized in the previous exercise.

11. Implement the circuit for chaos-based encryption described in Section 8.11.

Bibliography

[1] D. M. Abrams and S. H. Strogatz. Chimera states for coupled oscillators. *Physical Review Letters*, 93(17):174102, 2004.

[2] Arduino. Arduino guide. `https://www.arduino.cc/en/Guide/HomePage`, 2016. [Online; accessed 12 October 2016].

[3] D. P. Atherton. *Nonlinear control engineering.* Van Nostrand Reinhold, 1975.

[4] A. Bergner, M. Frasca, G. Sciuto, A. Buscarino, E. Ngamga, L. Fortuna, and J. Kurths. Remote synchronization in star networks. *Physical Review E*, 85(2):026208, 2012.

[5] S. Boccaletti. *The synchronized dynamics of complex systems.* Elsevier, 2008.

[6] S. Boccaletti, V. Latora, Y. Moreno, M. Chavez, and D.-U. Hwang. Complex networks: Structure and dynamics. *Physics Reports*, 424(4):175–308, 2006.

[7] Y. Braiman, J. F. Lindner, and W. L. Ditto. Taming spatiotemporal chaos with disorder. *Nature*, 378(6556):465, 1995.

[8] M. Bucolo, L. Fortuna, M. Frasca, and M. G. Xibilia. A generalized Chua cell for realizing any continuous n-segment piecewise-linear function. *International Journal of Bifurcation and Chaos*, 11:2517–2527, 2001.

[9] A. Buscarino, L. Fortuna, and M. Frasca. Chua's circuits synchronization with diffusive coupling. *International Journal of Bifurcation and Chaos*, 19:3101–3107, 2009.

[10] A. Buscarino, L. Fortuna, M. Frasca, L. V. Gambuzza, and G. Sciuto. Memristive chaotic circuits based on cellular nonlinear networks. *International Journal of Bifurcation and Chaos*, 22:1250070, 2012.

[11] A. Buscarino, L. Fortuna, M. Frasca, M. Iachello, and V.-T. Pham. Robustness to noise in synchronization of network motifs: Experimental results. *Chaos*, 22:043106, 2012.

[12] A. Buscarino, L. Fortuna, M. Frasca, and G. Sciuto. *A concise guide to chaotic electronic circuits.* Springer, 2014.

[13] A. Buscarino, L. V. Gambuzza, M. Porfiri, L. Fortuna, and M. Frasca. Robustness to noise in synchronization of complex networks. *Scientific Reports*, 3:2026, 2013.

[14] S. Camazine. *Self-organization in biological systems*. Princeton University Press, 2003.

[15] M. Candaten and S. Rinaldi. Peak-to-peak dynamics: A critical survey. *International Journal of Bifurcation and Chaos*, 10(08):1805–1819, 2000.

[16] A. J. Carlson, A. C. Ivy, L. R. Krasno, and A. H. Andrews. The physiology of free fall through the air: delayed parachute jumps. *Quarterly Bulletin of the Northwestern University Medical School*, 16(4):254, 1942.

[17] G. Chen and X. Dong. *From chaos to order: perspectives, methodologies and applications*. Singapore. World Scientific, 1998.

[18] A. Chiuso, L. Fortuna, M. Frasca, A. Rizzo, L. Schenato, and S. Zampieri. *Modelling, estimation and control of networked complex systems*. Springer, 2009.

[19] L. Chua. Memristor — the missing circuit element. *IEEE Transactions on Circuit Theory*, 18(5):507–519, 1971.

[20] L. O. Chua and T. Roska. *Cellular neural networks and visual computing: foundations and applications*. Cambridge University Press, Cambridge, 2002.

[21] L. O. Chua and L. Yang. Cellular neural networks: Theory. *IEEE Transactions on Circuits and Systems*, 35:1257–1272, 1988.

[22] P. A. Cook. *Nonlinear dynamical systems*. Prentice Hall, 1994.

[23] N. J. Corron. A simple circuit implementation of a van der Pol oscillator. `http://ccreweb.org/documents/physics/chaos/vdp2006.html`, 2010. [Online; accessed 12 October 2016].

[24] T. Deisboeck and J. Y. Kresh. *Complex systems science in biomedicine*. Springer Science & Business Media, 2007.

[25] R. Dogaru and L. O. Chua. Edge of chaos and local activity domain of FitzHugh–Nagumo equation. *International Journal of Bifurcation and Chaos*, 8:11–57, 1998.

[26] R. Dogaru and L. O. Chua. Edge of chaos and local activity domain of the Brusselator CNN. *International Journal of Bifurcation and Chaos*, 8:1107–1130, 1998.

[27] R. Dogaru and L. O. Chua. Edge of chaos and local activity domain of the Gierer-Meinhardt CNN. *International Journal of Bifurcation and Chaos*, 8:2321–2340, 1998.

[28] E. Estrada. *The structure of complex networks: theory and applications.* Oxford University Press, 2012.

[29] I. Fischer, R. Vicente, J. M. Buldú, M. Peil, C. R. Mirasso, M. Torrent, and J. García-Ojalvo. Zero-lag long-range synchronization via dynamical relaying. *Physical Review Letters*, 97(12):123902, 2006.

[30] L. Fortuna, M. Frasca, and A. Rizzo. Chaotic pulse position modulation to improve the efficiency of sonar sensors. *IEEE Transactions on Instrumentation and Measurement*, 52(6):1809–1814, 2003.

[31] L. Fortuna, M. Frasca, and M. G. Xibilia. *Chua's circuit implementations: yesterday, today and tomorrow.* World Scientific, 2009.

[32] L. Fortuna, A. Rizzo, and M. G. Xibilia. Modeling complex dynamics via extended PWL-based CNNs. *International Journal of Bifurcation and Chaos*, 13:3273–3286, 2003.

[33] L. V. Gambuzza, A. Cardillo, A. Fiasconaro, L. Fortuna, J. Gómez-Gardenes, and M. Frasca. Analysis of remote synchronization in complex networks. *Chaos: An Interdisciplinary Journal of Nonlinear Science*, 23(4):043103, 2013.

[34] L. Gammaitoni, P. Hänggi, P. Jung, and F. Marchesoni. Stochastic resonance. *Review of Modern Physics*, 70:223–287, 1998.

[35] R. Genesio and A. Tesi. Chaos prediction in nonlinear feedback systems. In *IEE Proceedings D-Control Theory and Applications*, volume 138, pages 313–320. IET, 1991.

[36] V. Govorukhin. Calculation of Lyapunov exponents for ODE. `https://it.mathworks.com/matlabcentral/fileexchange/4628-calculation-lyapunov-exponents-for-ode`, 2004. [Online; accessed 13 October 2016].

[37] J. Guckenheimer and P. J. Holmes. *Nonlinear oscillations, dynamical systems, and bifurcations of vector fields*, volume 42. Springer Science & Business Media, 2013.

[38] R. Gutiérrez, R. Sevilla-Escoboza, P. Piedrahita, C. Finke, U. Feudel, J. M. Buldu, G. Huerta-Cuellar, R. Jaimes-Reategui, Y. Moreno, and S. Boccaletti. Generalized synchronization in relay systems with instantaneous coupling. *Physical Review E*, 88(5):052908, 2013.

[39] H. Haken. *Synergetics: an introduction.* Springer-Verlagr, Berlin, 1989.

[40] L. Huang, Q. Chen, Y.-C. Lai, and L. M. Pecora. Generic behavior of master-stability functions in coupled nonlinear dynamical systems. *Physical Review E*, 80(3):036204, 2009.

[41] J. L. Huertas, W.-K. Chen, and R. N. Madan, editors. *Vision of nonlinear science in the 21st century*. World Scientific, Singapore, 1999.

[42] H. Kantz and T. Schreiber. *Nonlinear time series analysis*, volume 7. Cambridge University Press, 2004.

[43] D. Kaplan and L. Glass. *Understanding nonlinear dynamics*. Springer Science & Business Media, 2012.

[44] K. Karacs, G. Y. Cserey, A. Zarandy, P. Szolgay, C. S. Rekeczky, L. Kek, V. Szabo, G. Pazienza, and T. Roska. Software library for cellular wave computing engines. http://cnn-technology.itk.ppke.hu/Template_library_v3.1.pdf, 2010. [Online; accessed 12 October 2016].

[45] S. A. Kauffman. *The origins of order: self organization and selection in evolution*. Oxford University Press, USA, 1993.

[46] M. P. Kennedy. Three steps to chaos. I. Evolution. *IEEE Transactions on Circuits and Systems I: Fundamental Theory and Applications*, 40(10):640–656, 1993.

[47] M. P. Kennedy. Three steps to chaos. II. A Chua's circuit primer. *IEEE Transactions on Circuits and Systems I: Fundamental Theory and Applications*, 40(10):657–674, 1993.

[48] H. K. Khalil. *Nonlinear systems*. Pearson, 2001.

[49] Y. A. Kuznetsov. *Elements of applied bifurcation theory*, volume 112. Springer Science & Business Media, 2013.

[50] Y.-C. Lai and T. Tél. *Transient chaos: complex dynamics on finite time scales*, volume 173. Springer Science & Business Media, 2011.

[51] W. S. Levine. *The control handbook*. CRC Press, 1996.

[52] E. N. Lorenz. Deterministic nonperiodic flow. *Journal of the Atmospheric Sciences*, 20(2):130–141, 1963.

[53] D. G. Luenberger. Introduction to dynamic systems; theory, models, and applications. Technical report, 1979.

[54] R. N. Madan. *Chua's circuit: a paradigm for chaos*, volume 1. World Scientific, 1993.

[55] K. Mainzer. *Thinking in complexity: the computational dynamics of matter, mind, and mankind*. Springer Science & Business Media, 2007.

[56] G. Manganaro, P. Arena, and L. Fortuna. *Cellular neural networks: chaos, complexity and VLSI processing*. Springer, Berlin, 1999.

[57] J. E. Marsden and M. McCracken. *The Hopf bifurcation and its applications*, volume 19. Springer Science & Business Media, 2012.

[58] L. Min, K. R. Crounse, and L. O. Chua. Analytical criteria for local activity and applications to the Oregonator CNN. *International Journal of Bifurcation and Chaos*, 10:25–71, 2000.

[59] F. C. Moon. *Chaotic vibrations: an introduction for applied scientists and engineers*. New York, Wiley, 1987.

[60] J. D. Murray. *Mathematical biology I: an introduction, Vol. 17 of interdisciplinary applied mathematics*. Springer, New York, NY, USA,, 2002.

[61] M. Newman. *Networks: an introduction*. Oxford University Press, 2010.

[62] G. Nicolis. *Introduction to nonlinear science*. Cambridge University Press, 1995.

[63] G. Nicolis, I. Prigogine, and G. Nicolis. *Exploring complexity*. WH Freeman & Company, 1989.

[64] I. Omelchenko, Y. Maistrenko, P. Hövel, and E. Schöll. Loss of coherence in dynamical networks: spatial chaos and chimera states. *Physical Review Letters*, 106(23):234102, 2011.

[65] E. Ott, J. A. Yorke, and T. Sauer. *Coping with chaos: analysis of chaotic data and the exploitation of chaotic systems*. Wiley, 1994.

[66] M. J. Panaggio and D. M. Abrams. Chimera states: coexistence of coherence and incoherence in networks of coupled oscillators. *Nonlinearity*, 28(3):R67, 2015.

[67] T. S. Parker and L. Chua. *Practical numerical algorithms for chaotic systems*. Springer Science & Business Media, 2012.

[68] L. M. Pecora and T. L. Carroll. Synchronization in chaotic systems. *Physical Review Letters*, 64(8):821, 1990.

[69] L. M. Pecora and T. L. Carroll. Master stability functions for synchronized coupled systems. *Physical Review Letters*, 80(10):2109, 1998.

[70] H.-O. Peitgen, H. Jürgens, and D. Saupe. *Chaos and fractals: new frontiers of science*. Springer Science & Business Media, 2006.

[71] A. Pikovsky, M. Rosenblum, and J. Kurths. *Synchronization: a universal concept in nonlinear sciences*, volume 12. Cambridge University Press, 2003.

[72] W. H. Press, S. A. Teukolsky, W. T. Vetterling, and B. P. Flannery. *Numerical recipes in C*, volume 2. Cambridge University Press, 1996.

[73] O. E. Rössler. An equation for continuous chaos. *Physics Letters A*, 57(5):397–398, 1976.

[74] T. Saito. An approach toward higher dimensional hysteresis chaos generators. *IEEE Transactions on Circuits and Systems*, 37(3):399–409, 1990.

[75] E. Schöll and H. G. Schuster. *Handbook of chaos control*. John Wiley & Sons, 2008.

[76] T. Schreiber. Interdisciplinary application of nonlinear time series methods. *Physics Reports*, 308(1):1–64, 1999.

[77] D. Sherman. David Sherman Engineering Co. SoundArb. `http://slatecreekengineering.com/SoundArb.htm`, 2008. [Online; accessed 12 October 2016].

[78] J.-J. E. Slotine, W. Li, et al. *Applied nonlinear control*, volume 199. Prentice-Hall Englewood Cliffs, NJ, 1991.

[79] J. C. Sprott. *Chaos and time-series analysis*, volume 69. Oxford University Press, 2003.

[80] J. C. Sprott. *Elegant Chaos: Algebraically Simple Chaotic Flows*. World Scientific, Singapore, 2010.

[81] W.-H. Steeb. *A Handbook of Terms Used in Chaos and Quantum Chaos*. B.I. Wissenschaftsverlag, Berlin, 1991.

[82] L. Strauss. *Wave generation and shaping*. McGraw-Hill, 1970.

[83] S. Strogatz. *Sync: The emerging science of spontaneous order*. Hyperion, 2003.

[84] S. H. Strogatz. *Nonlinear dynamics and chaos: with applications to physics, biology, chemistry, and engineering*. Westview Press, 2001.

[85] D. B. Strukov, G. S. Snider, D. R. Stewart, and R. S. Williams. The missing memristor found. *Nature*, 453(7191):80–83, 2008.

[86] R. Thom. *Structural stability and morphogenesis: an outline of a general theory of models*. Addison-Wesley, 1989.

[87] C. Toumazou, S. Porta, and N. Battersby. *Circuits and Systems Tutorials: ISCAS' 94*. IEEE Press, 1995.

[88] J. J. Tyson. Modeling the cell division cycle: cdc2 and cyclin interactions. *Proceedings of the National Academy of Sciences*, 88(16):7328–7332, 1991.

[89] G. Vallis. Conceptual models of el nino and the southern oscillation. *Journal of Geophysical Research: Oceans*, 93(C11):13979–13991, 1988.

[90] B. Van der Pol. The nonlinear theory of electric oscillations. *Proceedings of the Institute of Radio Engineers*, 22(9):1051–1086, 1934.

[91] F. Varela, E. Thompson, and E. Rosch. *The embodied mind: cognitive science and human experience.* Cambridge, MIT Press, 1991.

[92] M. Vidyasagar. *Nonlinear systems analysis.* SIAM, 2002.

[93] G. M. Weinberg. *An introduction to general systems thinking.* Wiley, New York, 1975.

[94] J. C. Whitaker. *The electronics handbook.* CRC Press, 1996.

[95] E. C. Zeeman. Catastrophe theory. *Scientific American*, pages 65–70, 1976.

[96] E. C. Zeeman. *Catastrophe theory: selected papers, 1972–1977.* Addison-Wesley, 1977.

Index